Snakes of Virginia

SNAKES *of Virginia*

Donald W. Linzey

and

Michael J. Clifford

University Press of Virginia

Charlottesville and London

The publication of this volume is partially sponsored by the Wytheville Community College Educational Foundation, Inc.

University of Virginia Press

Printed in the United States of America on acid-free paper

9 8 7 6 5 4 3 2 1

First published 1981
First paperback edition published 2002
ISBN 0-8139-2154-6 (paper)

The Library of Congress has cataloged the hardcover edition as follows:

Library of Congress Cataloging in Publication Data
Linzey, Donald W.
Snakes of Virginia.
Bibliography: p. 143.
1. Snakes—Virginia—Identification. 2. Reptiles—Identification. 3. Reptiles—Virginia—Identification. I. Clifford, Michael J. II. Title.
QL666.06L74 597.96'09755 81-12951
ISBN 0-8139-0826-4 AACR2

Printed in the United States of America

TO OUR SONS

David, Tom, and Andy

PREFACE

The need for a layman's book on the snakes of Virginia has been recognized for many years. In April 1960 the Virginia Herpetological Society listed a color booklet on the amphibians and reptiles as one of its "long-range dreams." During the interim, the society under the very capable direction of Franklin J. Tobey, Jr., has issued mimeographed bulletins containing information useful for identifying Virginia's snakes as well as keeping its members informed about recent discoveries and range extensions in the state. Earlier, Tobey (1957) wrote a magazine article covering some of Virginia's harmless snakes. More recently, illustrated articles covering both nonpoisonous and poisonous snakes appeared in *Virginia Wildlife* (Mitchell, 1974*c*, 1794*d;* Mitchell and Martin, 1981).

In writing this book, we had several aims in mind. First, we wanted a book that could be read and understood by the layman and the nonscientist—the farmer, sportsman, game warden, hiker, or camper. Second, we wanted a book that could be used by young people and their leaders—4-H Club members, Boy Scouts, Girl Scouts, Cub Scouts, Brownies, YMCA groups, and city and county recreation departments. Third, our aim was to provide a source book for librarians, physicians, veterinarians, county and state extension personnel, and high school and college biology teachers. We feel that the text, drawings, and color photographs included in this book accomplish these aims and will meet the needs of each of these groups.

In addition, we have gathered together in one publication all known locality and range data on Virginia snakes for the use of both amateur and professional herpetologists. Many new locality records have been uncovered as a result of our search, and for the benefit of future researchers, a complete listing of all collections containing specific species is given at the end of each species account. An exhaustive search of the literature has also enabled us to compile a bibliography with over five hundred citations on the snakes of Virginia, beginning with an account from the year 1606 (Clarke, 1670). The result is a scientifically accurate book written in a form understandable to the nonscientist, yet containing data useful to the person with a deeper, more serious interest in the study of herpetology.

The destruction and degradation of habitat have been primarily responsible for the decline in numbers of the canebrake rattlesnake in

Virginia. Extensive land clearing, the destruction of natural stream and stream-associated habitats by channelization, the draining and filling of swamplands and marshes, the widespread use of environmental contaminants such as pesticides and other potentially dangerous chemicals—all of these are affecting Virginia's wildlife populations. However, the greatest number of snakes continue to meet their death by deliberate persecution because of the misinformation that abounds concerning them. It is our fervent hope that this book will help educate the misinformed and make people more aware of the beneficial aspects of this much maligned group of reptiles.

The idea for a book on the snakes of Virginia was born several years ago when Clifford began writing a series of species accounts. In 1977 Linzey and Clifford agreed to collaborate, combining data and talents and proceeding with publication. An intensive search of college, university, museum, and private collections was initiated in 1977.

It has taken the combined efforts of many persons to compile this publication. We wish to especially thank Steven Q. Croy for visiting a number of collections, verifying identifications, and recording data. In addition, his assistance in the field has been invaluable. Karen L. Moore has spent many hours in the library thoroughly searching state and national journals for records of Virginia specimens. She has ably and accurately transcribed field data and collection records and has assisted in the preparation of the final range maps. Franklin Tobey graciously agreed to review the manuscript and offered helpful suggestions.

We would like to extend our appreciation to the following individuals for either providing data and/or allowing us to examine specimens in their care: Fred J. Alsop III, East Tennessee State University; John S. Applegarth, University of New Mexico; Joseph R. Bailey, Duke University; Herbert Boschung, University of Alabama Museum of Natural History; Peter D. Bottjer, Yale University; Russell C. Brachman, Averett College; Garnett R. Brooks, College of William and Mary; Charles C. Carpenter, University of Oklahoma; Dennis L. Carter, Shenandoah National Park; Robert F. Clarke, Emporia State University; Kenneth Cliffer, University of Minnesota; John M. Condit, Ohio State University; Costello M. Craig, Bedford, Virginia; Deborah Dattner, Peninsula Nature and Science Center; Mark Dodero, San Diego Society of Natural History; Neil H. Douglas, Northeast Louisiana University; Ralph P. Eckerlin, Northern Virginia Community College; Carl H. Ernst, George Mason University; Franklin F. Flint, Randolph-Macon Woman's College; Paul Gritis, Field Museum of Natural History; Frank Groves, Baltimore Zoo; Herbert S. Harris, Jr., Natural History Society of Maryland; Marvin M. Hensley, Michigan State University; Ronald Heyer, National Museum of Natural History; Richard Highton, University of Maryland; Edmond V. Malnate, Academy of Natural Sciences, Philadelphia; M. Elizabeth McGhee, University of Michigan; Joseph F. Merritt, Old Dominion University; Michael A. Morris, Illinois State Natural History Survey; John A. Musick, Virginia Institute of Marine Science; Marc R. Nadeau, University of Kansas; Douglas W. Ogle, Virginia Highlands Community College; William M. Palmer, North Carolina State Museum of Natural History; F. Harvey Pough, Cornell University; Jose P. Rosado, Harvard University; Robert D. Ross, Virginia Polytechnic Institute and State University; Douglas A. Rossman, Louisiana State University; Albert E. Sanders, The Charleston Museum; Donald J. Schwab, Suffolk, Virginia; William H. Stickel, Patuxent Wildlife Research Center; Grace M. Tilger, American Museum of Natural History; Arlene L. Webb, University of Illinois; Alan G. C. White, Virginia Military Institute; Shirley Whitt, Lynchburg College; and Gary M. Williamson, Norfolk, Virginia.

The following individuals either searched their collections and reported having no specimens from Virginia or assisted in data gathering in other ways: Alexander J. Barton, National Science Foundation; Joseph F. Beckler, Toledo Zoo; Bob Brown, University of Louisville; Clare E. Close, University of Georgia; Donald E. Hahn, Cottonwood, Arizona; William F. Harrington, Johns Hopkins University; Frances C. James, Florida State University; Lynne Kunze, Virginia Western Community College; Michael L. McMahan, Campbellsville College; Les Meade, Morehead State University; Jerry Nagel, East Tennessee State University; and Michael E. Seidel, Marshall University.

The line drawings and maps were prepared by Georgia Minnich. Photographs have been secured from Roger Barbour, Steven Croy, James Hill, Zygmund Leszczynski, Karl Maslowski, John MacGregor, Robert Mount, and A. Floyd Scott.

In addition to those persons recognized in the first printing for providing data, we would like to add David Auth, Florida Museum of Natural History; Elyse Beldon, National Museum of Natural History; Paisley Cato, Virginia Museum of Natural History; Ellen Censky, Carnegie Museum; Greg Schneider, University of Michigan Museum of Zoology; and Shi-kuei Wu, University of Colorado Museum. We also wish to thank Franklin Tobey for his helpful comments and clarification of some of the records in his distributional survey (1985).

Finally, we wish to express our appreciation to the Wytheville Community College Educational Foundation, Inc. for their financial assistance in the preparation and publication of the third printing.

We have made every attempt to gather all known existing data on Virginia's snakes. However, in any effort of this magnitude, some data may have been overlooked. We request that any individual with additional information please get in contact with us so that records may be updated and kept current.

DONALD W. LINZEY
Department of Biology
Wytheville Community College
Wytheville, Virginia 24382

MICHAEL J. CLIFFORD
Cooperative Extension Service
Virginia Polytechnic Institute and State University
Nottoway, Virginia 23955

CONTENTS

COLOR PLATES

Snakes of Virginia

INTRODUCTION

Snakes are among the most fascinating of all animals. They seem strange and mysterious with their odd shape and gliding movement, their silent lives and flicking tongue, and because some are poisonous, many people are deathly afraid of them all. For creatures without legs, their many successful adaptations are really logical ways of living. Although feared by many persons and even killed by some, these unique reptiles play an important role in the biological world.

All snakes are classified as vertebrates—animals that possess a vertebral column, or backbone, composed of individual segments known as vertebrae. Other vertebrates include fish, salamanders, frogs, turtles, lizards, alligators and crocodiles, birds, and mammals. Since humans are mammals, they are also classified as vertebrates.

Certain of these animals—the snakes, turtles, lizards, and crocodilians—share certain characteristics. These animals, the reptiles, are thus placed in the class Reptilia. The most obvious external feature of reptiles is the fact that their bodies are covered by dry scales. This is in contrast to the smooth skin of amphibians, the feathers of birds, and the hair of mammals. In addition, most reptiles have a three-chambered heart, an advanced type of kidney, very few integumentary glands and are cold-blooded. This latter term does not mean that the blood of reptiles is always cold. Rather, it means that reptiles cannot internally regulate their body temperature. The body temperature of a warm-blooded animal such as a bird or mammal remains relatively constant regardless of the surrounding air temperature. The body temperature of a reptile, on the other hand, fluctuates with the surrounding air temperature. When the temperature drops, so does the reptile's body temperature, and the animal becomes sluggish. Snakes have to keep their body temperature above 55°F in order to move about. If their temperature goes much below 40°F, they become stiff and cannot move. When the temperature rises, the reptile becomes active again. It is even possible for the body temperature of a cold-blooded reptile to be higher than the body temperature of a warm-blooded bird or mammal. This situation might arise when a snake suns itself on a warm rock or paved road, although such temperatures cannot be tolerated for very long.

Snakes, in turn, possess several features that differentiate them from all of the other reptiles. The elongate body and the absence of limbs

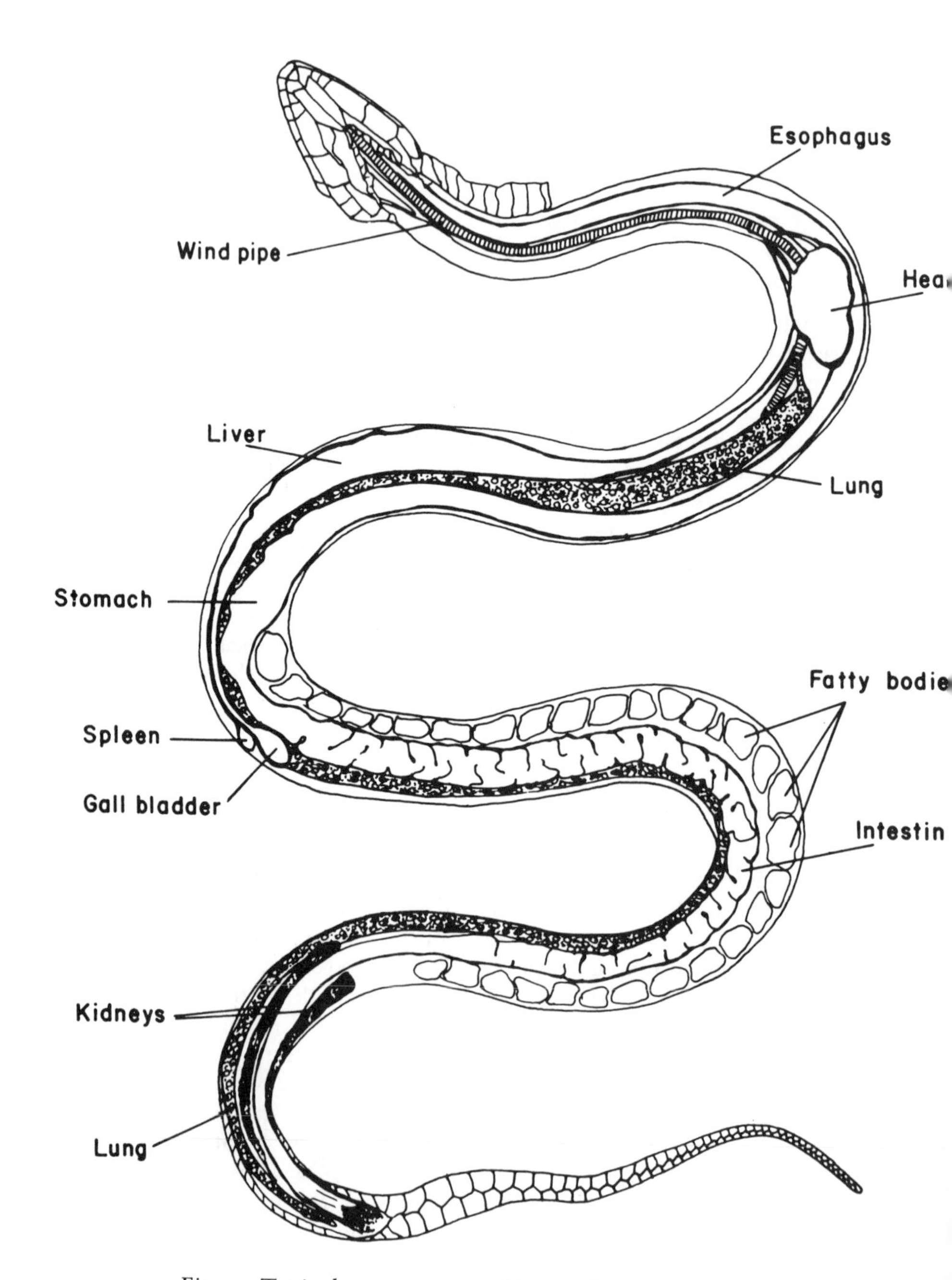

Fig. 1. Typical arrangement of internal organs in a snake.

are the most obvious features. In addition, snakes lack movable eyelids and external ear openings. Internally, snakes lack a sternum (breastbone), a eustachian tube, and a urinary bladder. Many snakes possess only a single functional lung, which may be almost as long as the body. In addition, many of the internal organs, including the heart, are elongate in shape in order to conform to the elongate body (fig. 1).

Scales

The bodies of all snakes are covered by dry epidermal scales. Each scale projects backwards to partially overlap the one behind it. The number, shape, and size of scales around the head are important features in identifying the different kinds of snakes (fig. 2). These head scales are given specific names. For example, the scales around the eyes are the oculars, the scales surrounding the nasal openings are the nasals, and the scales along the edge of the lower lip are the lower labials. The kinds of scales on the dorsal surface of the body together with the number of scales per row are also useful in classification (fig. 3). The dorsal scales on some snakes have a longitudinal ridge, or keel, running the length of each scale (fig. 4). These scales are referred to as being *keeled,* and the snake often has a roughened appearance. Snakes that lack keels are said

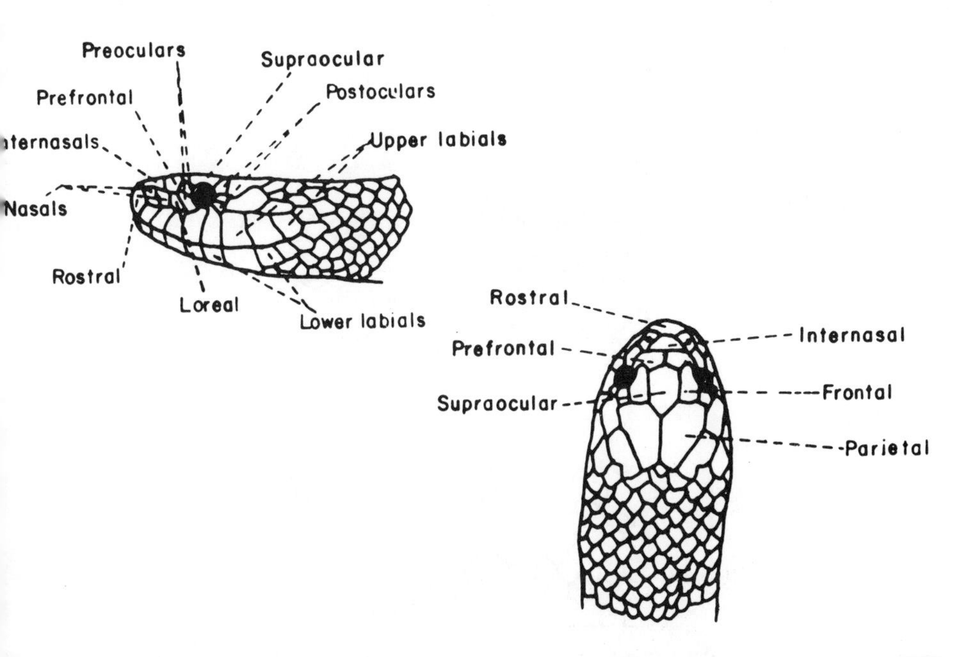

Fig. 2. Lateral and dorsal views of typical snake head showing locations of scales

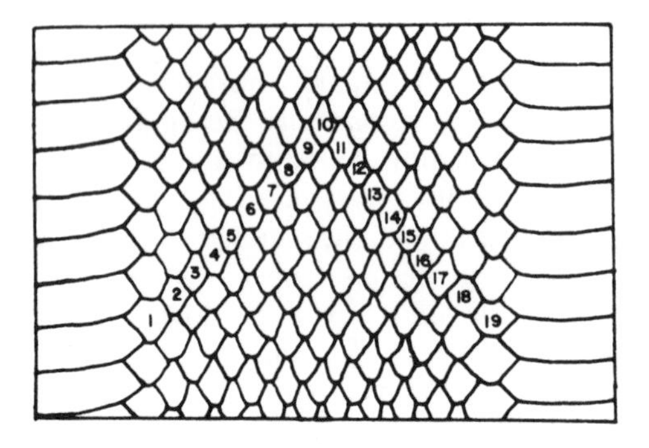

Fig. 3. Dorsal view of snake showing proper method of making a scale count

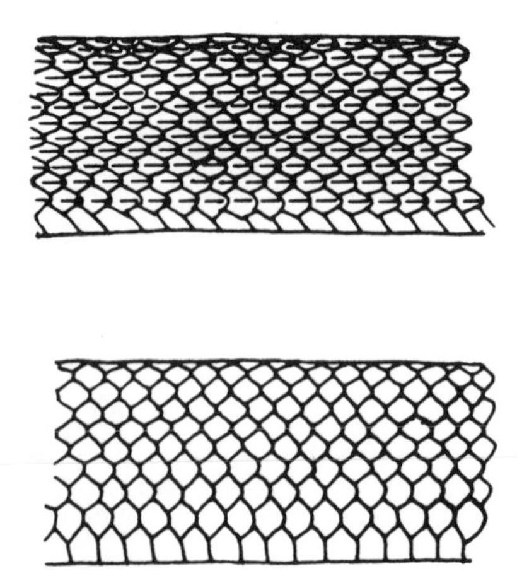

Fig. 4. Upper: *Keeled scales.* Lower: *Smooth (unkeeled) scales*

to have *smooth* scales. The belly is covered with transversely widened scales known as ventrals, a modification which makes them useful in locomotion. The transverse scales covering the bottom of the tail are known as caudals and may be either paired or unpaired (fig. 5). The cloacal opening, or vent, is covered by the anal scale or plate. This scale may exist as a single structure (*undivided anal plate*), or it may be divided into two parts (*divided anal plate*). The condition of the anal scale serves as a useful tool in classification. The presence or absence of various types of pigment-bearing cells (*chromatophores*) in the scales determines the color pattern of a snake. Thus, the number, kind, arrangement, and color of the various scales on the body of a snake not only give rise to the distinctive pattern of each species but are useful in keying out and identifying an unknown kind of snake.

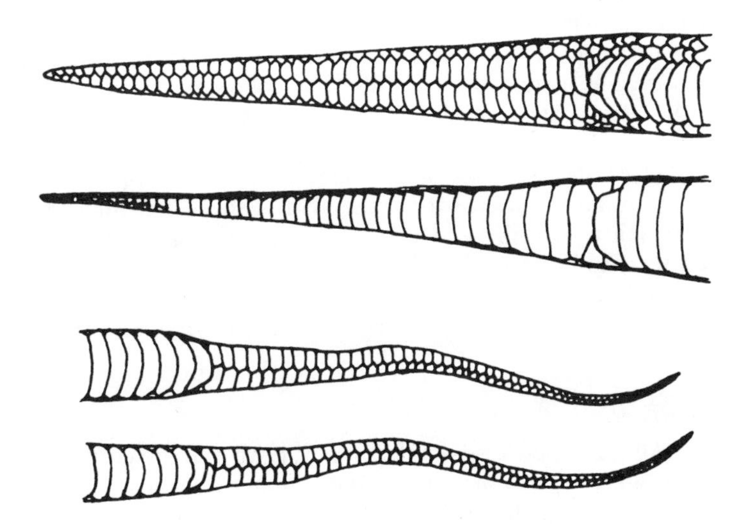

Fig. 5. Upper: *Ventral view of tail showing paired caudal scales and divided anal plate.* Lower: *Ventral view of tail showing unpaired caudal scales and undivided anal plate.*

As snakes grow they shed the outer layer of their skin. The frequency of shedding depends upon the age and rate of growth of the individual snake. A growing snake usually sheds about every two months, whereas a large adult snake may shed its skin only once a year. Most newborn snakes generally shed for the first time within 10 days following birth or hatching, although some shed within 36 hours of their birth.

Prior to shedding, the snake grows a new layer of epidermis beneath the outermost layer. About a week or so before the snake sheds, its pattern and color become dull and the scales covering the eyes turn an opaque milky white. At this time the snake is partially blind and often goes into hiding. Two or three days before the actual molting begins, the outermost layer of skin loosens and the eyes clear up so that the snake looks almost normal. Shedding begins when the snake rubs its nose and facial region against a log, rock, or some other roughened surface, causing a break in the old skin. By rubbing, the snake causes the skin to peel back over its head as well as under its chin. Once this occurs, the normal crawling movements of the snake allow it to literally crawl out of its old skin. The old skin peels back like a glove being turned inside out. The snake secretes an oily solution under the old skin which lubricates it and makes it easier for the snake to slip the old skin off. A healthy snake will normally shed its skin in one piece. The recently shed old skin looks like a piece of very thin plastic and feels soft and pliable. Every minute detail of the outer surface of the snake including the head, eye, lip, chin, belly, anal, and tail scales is visible. In fact, if a complete snake skin is found in the field, it is often possible to use it together with a key to identify exactly what type of snake it same from. The new skin is shiny, bright, and slightly moist, but it soon becomes dry. The colors are bright and fresh because the new scales are clear and have not yet become scratched and worn.

In most snakes the skin is shed down to the tip of the tail. In rattlesnakes, however, each time the skin is shed, a remnant of the old skin remains near the end of the tail and forms a new segment of the rattle. Thus, the rattle consists of a series of dry, horny, interlocking, loosely connected segments of epidermis. The rattling sound is produced by the segments striking against each other when the tail is vibrated.

Teeth

All snakes possess simple, conical teeth that are curved posteriorly. Teeth are present not only on the jawbones but also usually on one or more bones in the roof of the mouth. The number of teeth in snakes is highly variable even among members of the same species. Teeth frequently become damaged or broken, but this does not cause much of a problem because a new tooth will grow in to replace the broken or damaged tooth. If this new tooth should in turn be damaged, another tooth will replace it. Thus, snakes have a succession of teeth during their lifetime, rather than just two sets as do most mammals.

The hollow fangs of our poisonous snakes are modified teeth and are located on the anterior portion of the upper jaw. Like the other teeth, the fangs are replaced if they are lost or damaged.

A few of our snakes, including the hognose snake and the crowned snake, have several enlarged teeth at the rear of their jaws. Although the saliva of the hognose snake has been found to be toxic to frogs and toads, it is only mildly toxic to humans. The enlarged rear teeth do not function as fangs. They are used primarily for puncturing inflated toads and for manipulating food materials for swallowing. The enlarged teeth of the crowned snake are somewhat similar to the rear fangs found in some snakes native to other areas of the United States and to other parts of the world. Grooves are present along the surface of these enlarged teeth, but no poison glands are present in the crowned snake.

Senses

Snakes have very good vision for short distances and some can see very well at night. They have no eyelids and therefore cannot blink. Their eyes are protected by a transparent skinlike covering, or eye scale.

Snakes have a keen sense of smell. The sense of smell is heightened by the use of the forked tip of the tongue which "tastes" the air, picking up chemicals and small particles and conveying them to two small saclike structures that are located toward the front of the roof of the mouth. Each sac has a short duct or tube which opens into the mouth. These structures comprise the vomeronasal, or Jacobson's organ. To sense its surroundings, the snake flicks out its tongue, which picks up particles in the air or on the ground. Then the snake withdraws the tongue into its mouth and inserts the two tips into the two ducts. The Jacobson's organ then sends information regarding the habitat, potential food, mates and/ or enemies to the brain. The tongue is in no way poisonous or harmful, even in venomous species. Taste buds are lacking in snakes.

All snakes are extremely sensitive to vibrations that they receive through the ground or the substrate on which they are resting. Snakes lack a middle ear and an eardrum and are probably insensitive to most sounds. However, Wever and Vernon (1960) demonstrated that at least some species of colubrid snakes can hear sound waves traveling through the air as humans do.

One group of snakes, the pit vipers, possesses a pair of heat-sensing pits on each side of the head between the eye and the nostril. These sensory devices are discussed more fully on pages 123–124.

Reproduction

Most snakes breed once a year, with mating usually occurring in the spring or early summer. During this time, female snakes secrete a strong odor from their skin and from glands at the base of their tails. This odor leaves a trail that male snakes can easily follow. Fertilization is internal, with the males having special paired copulatory organs known as hemipenes, one of which conducts the sperm into the female's reproductive tract. The young either hatch or are born during the summer or early fall. The young are given no parental care after birth.

Snakes bring forth their young in different ways. Some snakes deposit leathery-shelled eggs in the soil, beneath piles of leaves, or in rotting logs and stumps. The developing embryo and its yolk supply are contained within the egg. The incubation period is variable, depending mainly upon temperature, but usually ranges between two and three months. Such snakes are said to be *oviparous.* Females of other species retain the young within the uterus until the time of birth. Each developing snake is enclosed in its own membrane and nourished by its own supply of yolk. Either just before or just after birth, the young snake breaks out of the membrane surrounding it. This method of development is known as *ovoviviparous* development. A few snakes retain the young within their uterus and actually nourish them through a placentalike structure until they are ready to be born. These snakes are referred to as *viviparous* species. By basking and absorbing heat, ovoviviparous species gain an advantage in reproduction in cooler areas over strictly oviparous forms. The chances for the young to survive are generally better with live births than with unprotected eggs.

Newborn snakes in Virginia vary in size from approximately 4 to 18 inches in length. Their pattern is usually similar to the parents, but not always. Young black racers, for example, are marked with dark gray blotches on a light gray background, while the adult is a uniform black with no pattern.

Unlike birds and mammals which grow to a specific size and then stop growing, snakes continue to grow throughout life. Growth is most rapid in the young and slower in older individuals. Snakes become reproductively mature before they are fully grown. Thus, sexual maturity seems to be associated more with the snake's attaining a certain size than with its attaining a specific age.

External differences between males and females are not readily apparent in most snake species. Male snakes normally have somewhat longer tails than females of the same species. In addition, the base of the tail in a male snake will normally appear wider or stouter than in a female snake of the same species due to the presence of the retracted paired hemipenes, or copulatory organs.

Food

All snakes are carnivorous. Most of the smaller snakes eat worms, insects, insect larvae, snails, millipedes, and centipedes. Larger snakes feed primarily on fish, salamanders, frogs, toads, lizards, reptile eggs, birds and their eggs, and small mammals. The eggs and young of ground-nesting birds are more vulnerable than are the eggs and young of tree-nesting or hole-nesting birds, although snakes such as the corn snake and black rat snake are excellent climbers and are able to reach nests in any location. Most of the small mammals consumed by snakes consist of mice, rats, and shrews, although chipmunks, squirrels, and rabbits are regularly taken by some of our larger snakes. Most snakes have a varied diet, but some have definite food preferences. For example, hognose snakes prefer toads, mud snakes prefer amphiumas, rainbow snakes prefer eels, and kingsnakes are well known for feeding upon other snakes, including poisonous species.

Since the teeth of snakes are pointed and curved backwards, they cannot use them to chew and, therefore, must eat their prey all in one piece. Most prey is swallowed headfirst.

Some snakes simply seize their prey and swallow it, although they sometimes throw a coil of their body over a particularly vigorous victim for added control. Others, such as the corn snake and black rat snake, are constrictors. Constrictors kill their prey by constricting or squeezing their bodies around the animal so that it cannot breathe. Pine snakes are also constrictors, but instead of wrapping their bodies around their prey, they use their powerful muscular body to squeeze the prey against the side of their burrow or their cage until suffocation occurs. Poisonous snakes use venom to subdue their prey.

Snakes have several adaptations for eating animals that are larger than the snake's normal mouth size. The mouth of most snakes extends back beyond the eyes to allow for a wide opening. The two halves of the lower jaw are not fused together, but are connected by an elastic ligament. This allows each half of the lower jaw to move independently. These bones can move forward, backward, up, down, and out to the sides. In addition, the tooth-bearing bones of the upper jaw are also able to move independently. Furthermore, each lower jaw attaches to the skull by means of two movable bones that give additional flexibility. In reality, the snake slowly crawls or "walks" itself forward over the prey, rather than swallowing it as a mammal does. The glottis of a snake is adapted in such a way that it can be protruded so that the air passage is kept clear while large prey is being slowly swallowed.

Habitats

Snakes inhabiting Virginia may live primarily on land (terrestrial) or divide their time between land and water (semiaquatic). Most of the terrestrial snakes live on the surface of the ground (garter snake, brown snake, hognose snake), although some may be mainly arboreal (rough green snake) and others are mainly fossorial (earth snake, mole snake, crowned snake). Semiaquatic species are most frequently encountered in freshwater streams, lakes, or swamps, although some venture into brackish water habitats.

A few of our snakes live in very restricted ecological habitats (mud snake, rainbow snake), whereas others may range from coastal areas to the mountains (garter snake, hognose snake, copperhead). Habitat destruction and pollution as a result of human activities are probably the worst enemies of snake populations. The draining and filling of swamps and marshes destroy countless acres of habitat every year. The pollution of streams with industrial wastes and organic sewage make these areas unfit for many forms of life. As new shopping centers and housing developments continue to be constructed, animal populations are forced into smaller and smaller fragments of suitable habitat. In many areas this has resulted in the disappearance of certain forms.

Locomotion

Virginia's snakes utilize three different methods of locomotion. Most of our snakes move in a series of waves of S-shaped curves, with each part of the body passing along the same track. Strong muscles attached to the sides of the vertebrae alternately contract and relax to bend the spine. Snakes can get enough traction to thrust their bodies forward by pushing against a projection such as a clump of grass, a pebble, an exposed rock, or just the irregular surface of the ground. This serpentine, or undulatory, movement may be used on the ground, while swimming, or when the snakes are climbing in bushes and trees.

In certain situations, snakes can use their belly or ventral scales to alternately push and pull themselves along in a straight line. The ventral scales protrude slightly and their free edges are able to move forward and backward, thus they are able to "grip" the ground like tractor treads. The snake raises its ventral scales, moves them forward a short distance, and then allows them to become anchored to the ground or substrate. Then the body literally slides forward until it becomes reoriented in the skin. Progress is effected by a series of such movements, with the snake appearing to move in a straight line without any lateral motion, somewhat similar to a crawling caterpillar.

A third method of crawling is by bunching up the S-shaped curves so that they lie close together like the bellows of an accordion. Movement consists of alternately curving and straightening the body. After the body is drawn up in a series of curves, the tail serves as an anchor and the rest of the body is pushed forward until it is straight. The weight and friction of the bunched curves give the snake enough traction so that it can thrust its head forward. The head and neck then serve as anchors, while the rest of the body is drawn up in a series of curves.

Hibernation

Falling temperatures during the autumn cause snakes and other cold-blooded vertebrates to seek a hibernaculum, or winter resting place. Animal burrows, rock crevices, caves, mine tunnels, and even deserted anthills afford snakes suitable overwintering sites below the frost line. Individuals of some species begin congregating at communal den sites in late summer. The presence of available wintering sites may be a critical factor controlling the abundance of snakes in certain regions.

Numerous instances have been recorded of more than one species occupying a particular den. One ant mound in Minnesota contained 148 smooth green snakes, 101 red-belly snakes, and 8 Plains garter snakes (Criddle, 1937). A hibernaculum in a gravel bank in Pennsylvania contained 16 snakes of 7 different species plus four salamanders

(Lachner, 1942). An ant mound in Michigan was excavated to a depth of 30 inches. Of the 77 vertebrates taken from the mound, 62 were snakes representing 7 different species and 15 were amphibians of 3 species (Carpenter, 1953). Canadian zoologists observed 10,000–15,000 red-sided garter snakes emerging from hibernation from one snake pit in Manitoba (Anon., 1978). Studies have shown that individual snakes often return to the same den to hibernate year after year.

Defensive Behavior

Many persons believe that all snakes are aggressive. Stories are even told about snakes chasing people. Except for certain snakes such as the cottonmouth and rattlesnake that often stand their ground when disturbed, most of our other snakes will do everything they can to get away from an intruder or a potential enemy. If they find themselves cornered, however, and escape is not possible, most snakes will resort to striking and biting in order to defend themselves. Pine snakes hiss loudly, in addition to striking, whenever disturbed. Hognose snakes attempt to intimidate potential enemies by flattening their neck, hissing, and, if necessary, playing dead. Many of our snakes become nervous when disturbed and vibrate their tails. If they happen to be in dry leaves at such times, the sound is very similar to the buzzing produced by the rattles of a rattlesnake.

Certain snakes, especially our nonpoisonous water snakes, garter snakes, and rat snakes, are well known for their habit of thrashing about when captured and smearing the contents of their anal scent glands on their captor. These scent glands are located near the base of the tail and the fluid is discharged through the vent. Even though it exits via the vent, the fluid does not come from the digestive system and has no relationship with the feces of the snake. The secretion, known as musk, is usually yellowish or whitish and has a very penetrating, disagreeable, musky odor.

Predation and Mortality

Snake eggs are eaten by opossums, raccoons, weasels, skunks, and other snakes. Small adult snakes as well as juvenile snakes may fall prey to fish, turtles, other snakes, herons, egrets, ibis, crows, hawks, owls, and various mammals. Larger snakes have fewer predators. Man is a major enemy of snakes, both small and large. An irrational fear of these animals, largely born of ignorance, results in large numbers of snakes being killed on sight. Overcollecting by amateur and professional herpetologists may be a factor in the decline of some species. Many captured snakes die in captivity due to lack of proper care. Roadside "snake-pit" menageries should be discouraged or strictly regulated as these often represent death traps for animals not properly cared for. Large numbers have probably fallen victim to persistent pesticides such as DDT, heptachlor, and dieldrin. Snakes, like other animals, are often killed while crossing roads and highways. Fatalities are increased due to the fact that many of these cold-blooded animals seek the warmth of road surfaces at night to warm their bodies.

Ecological Importance

Snakes play an important role in the biological world. Many kinds feed on mice and rats, thus keeping the populations of these small mammals under control. Corn snakes and rat snakes are especially beneficial in this respect, and many farmers encourage their presence around their barns. Studies have determined that a single rat can cause from $1.20 to over $20.00 in damage per year on farms (Dudderar and Beck, 1973). Thus, the presence of rodent-eating snakes on a farm can prevent hundreds or even thousands of dollars in damages. These snakes may occasionally consume birds or bird eggs, but the good they do far outweighs the bad.

Snakes and their eggs serve as food for other animals. Opossums, raccoons, mink, and weasels are fond of snake eggs. Snakes are known to be fed upon by some fish, turtles, other snakes, herons, egrets, cranes, cormorants, hawks, owls, crows, jays, opossums, raccoons, otters, foxes, and bobcats. Even bullfrogs occasionally eat small snakes. Rattlesnake meat is considered a delicacy by some humans.

Certain snakes are of great importance to medical science. The venom of poisonous snakes is used in the manufacture of antivenins used in the treatment of poisonous snakebite. Cottonmouth venom has been used as an anticoagulant for treating hemophilia. Enzymes from the snake's venom help form clots that stop the bleeding. Enzymes from vipers that inhabit other parts of the world, such as Russell's viper, are also used in helping to treat hemophilia. On the other hand, an enzyme taken from another viper's poison helps dissolve blood clots which can cause heart attacks and strokes.

Endangered and Threatened Species

The federal list of Threatened and Endangered Species in the United States does not include any species or subspecies found in Virginia. An unpublished master's thesis by Russ (1973) classified the pine snake, scarlet kingsnake, and canebrake rattlesnake as endangered in Virginia and the brown water snake, red-belly water snake, rainbow snake, mud snake, coastal plain milk snake, southeastern crowned snake, and eastern cottonmouth as rare. Such a classification for most of these endangered and rare species resulted from the fact that they reach the northernmost terminus of their range in Virginia.

In May 1978 a statewide Symposium on Endangered and Threatened Plants and Animals of Virginia was sponsored by the Center for Environmental Studies at Virginia Polytechnic Institute and State University in Blacksburg, Virginia (Linzey, 1979). The Committee on Amphibians and Reptiles, chaired by Franklin J. Tobey, Jr., made the following recommendations:

1. No snakes were classified as endangered or threatened.

2. The canebrake rattlesnake was classified as "Of Special Concern."

3. The eastern glossy water snake, the northern pine snake, and the southeastern crowned snake were classified as "Status Undetermined" due to the fact that very little information is available concerning their ranges in Virginia.

A second statewide symposium was held in 1989 (Terwilliger, 1991). The Committee on Amphibians and Reptiles, chaired by Joseph C. Mitchell, recommended:

1. The canebrake rattlesnake "be added to the state endangered species list because of the high rate of habitat loss and potential extirpation in the portion of its range encompassed by Virginia."

2. The mountain earth snake be considered for special concern status.

3. The glossy crayfish snake, smooth green snake, northern pine snake, black kingsnake, and southeastern crowned snake be classified as status undetermined due to inadequate information with which to make an informed decision about their status.

Care in Captivity

Basic suggestions for caring for captive snakes are given in most of the species accounts. Cages should be constructed of either wood and/or glass. Wire should be avoided as much as possible since many snakes have a tendency to rub their noses raw against roughened surfaces. Pegboard may be substituted for wire to add ventilation.

A glass aquarium with a wood and wire or pegboard lid is adequate for many species (fig. 6). Weights should be kept on the lid if the snake or snakes inside are strong enough to push the lid off.

Larger specimens are usually kept in wooden cages that have a glass front and a hinged top made of hardware cloth (see fig. 6). Most of these cages should be equipped with a hasp and lock to prevent other persons from handling the snakes and possibly allowing the snakes to escape. Cages that open from the top rather than from the front permit the handler to have much greater control over the animals in the cage.

The floors of most cages should be covered with newspaper or paper toweling to facilitate cleaning. Clean cages are essential for maintaining healthy animals. A container of fresh water—of a size suitable for the size and kind of snake in the cage—should be present at all times. Some snakes require water only for drinking, while others need a container large enough in which to partially or completely submerge their body. All cages should be equipped with some sort of shelter. For small snakes, pieces of bark or dried leaves will provide sufficient cover. Larger snakes will need a cardboard or wooden box into which they can retreat. A tree branch may be included in cages containing snakes that like to climb. When snakes in cages are preparing to shed, it is helpful to provide them with a roughened surface such as a rock, a log, or a large pinecone to use in the initial stages of the shedding process. A regular feeding schedule in which appropriate foods are provided should be established and maintained.

Most persons should not attempt to maintain poisonous snakes in captivity, particularly in a private home or apartment. In places

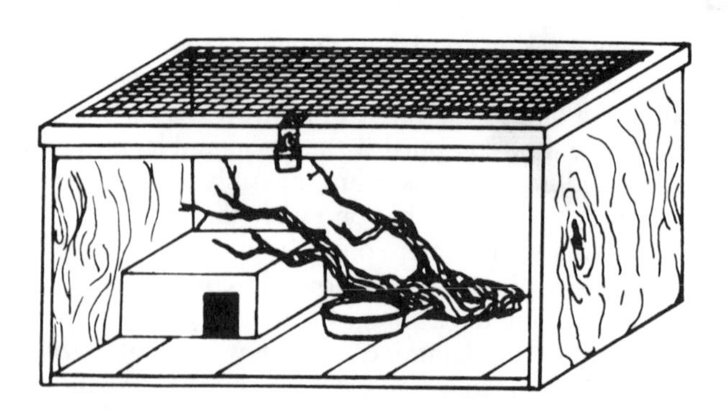

Fig. 6. Cage for small snakes (top) *and large snakes* (bottom)

where venomous snakes are maintained—zoos, research centers, universities—extreme caution must be exercised at all times.

Organizations

For the person with a serious interest in herpetology, there exist three professional societies in the United States:

Herpetologists' League
  Publishers of *Herpetologica*
The American Society of Ichthyologists and Herpetologists
  Publishers of *Copeia*
Society for the Study of Amphibians and Reptiles
  Publishers of *Journal of Herpetology and Herpetological Review*

Information concerning membership and activities of these organizations as well as the current addresses of their officers may be found in recent issues of their publications.

In Virginia, persons with a strong interest in reptiles and amphibians may wish to join the Virginia Herpetological Society. This is a nonprofit tax exempt organization devoted to increasing people's knowledge of Virginia's amphibians and reptiles and to encouraging broader public understanding and conservation of these animals. The society issues bulletins several times each year. For further information, contact Bob Hogan, Secretary/Treasurer, P.O. Box 603, Troutville, Virginia 24175.

A pamphlet entitled *Herpetology as a Career,* which describes opportunities and academic preparation for a career in herpetology, is available from the Publications Secretary of the Society for the Study of Amphibians and Reptiles.

Classification

The specific name of a snake consists of the generic name plus the trivial, or species, name. Members of a species breed with each other but normally not with other species. If a species is found to comprise two or more geographic races, each is given a third, or subspecific, name. A subspecies represents a part of the total population of a species; it occupies a definite geographic area and interbreeds with other populations of the species where they meet. The name of the person who first recognized the particular snake and gave the animal either its specific or subspecific name often follows the scientific name. Sometimes the author's name is in parentheses, as for example, *Farancia abacura* (Holbrook). This indicates that when Holbrook initially named the species, it was known by a different generic name.

In order to illustrate the complete classification of a snake, let us look at the black kingsnake. This animal would be classified as:

Kingdom	Animal
Phylum	Chordata
Subphylum	Vertebrata
Superclass	Tetrapoda
Class	Reptilia
Order	Squamata
Suborder	Serpentes
Family	Colubridae
Genus	*Lampropeltis*
species	*getulus*
subspecies	*niger*

The timber rattlesnake, however, would be classified in the following manner:

Kingdom	Animal
Phylum	Chordata
Subphylum	Vertebrata
Superclass	Tetrapoda
Class	Reptilia
Order	Squamata
Suborder	Serpentes
Family	Viperidae
Genus	*Crotalus*
species	*horridus*
subspecies	*horridus*

Species Accounts

An individual species account has been prepared for each of the 30 species of snakes currently known to inhabit Virginia. Each account gives the following information: common names, a brief description of the species, habitat, range in the United States and in Virginia, habits, reproduction, longevity, food, enemies, care in captivity, folklore, and location of specimens. Longevity data are taken from Bowler (1977) and are primarily from specimens maintained in captivity. The figures reflect the maximum life span of the species or subspecies wherever it was maintained and are not restricted just to Virginia specimens. A range map has been prepared for each species showing the counties in which it has actually been taken. All records have been verified by Linzey, Tobey (1985), and/or by the Virginia Herpetological Society. Common and scientific names follow the recommendations of Collins et al. (1990).

Location of Specimens

Each museum, college, university, or private collection originally containing one or more Virginia specimens and/or records is listed. Some private collections have been donated to larger institutions in order that they may be adequately maintained. Abbreviations used are:

AMNH American Museum of Natural History, New York City
ANSP Academy of Natural Sciences of Philadelphia
ART Private collection of A. R. Turner, Portsmouth, Virginia
ASU Arizona State University, Tempe
BC Bridgewater College, Bridgewater, Virginia
CAS Chicago Academy of Science
CHE Private collection of Carl H. Ernst, Fairfax, Virginia
CHM The Charleston Museum, Charleston, South Carolina
CM Carnegie Museum, Pittsburgh
CU Cornell University, Langmuir Laboratory, Ithaca, New York
CUC Columbia Union College, Takoma Park, Maryland
CWM College of William and Mary, Williamsburg, Virginia
DEH Private collection of Donald E. Hahn, Cottonwood, Arizona
DS Private collection of Donald Schwab, Suffolk, Virginia
DU Duke University, Durham, North Carolina
DWL Private collection of Donald W. Linzey, Blacksburg, Virginia
EH Emory and Henry College, Emory, Virginia
ESU Emporia State University, Emporia, Kansas
FMNH Field Museum of Natural History, Chicago
FSU Florida State University, Tallahassee
GDS Great Dismal Swamp Collection, Suffolk, Virginia (R. H. Rageot Collection)
GMU George Mason University, Fairfax, Virginia
GMW Private collection of Gary M. Williamson, Norfolk, Virginia
HSH Private collection of Herbert S. Harris, Baltimore, Maryland

INHS Illinois Natural History Survey, Urbana
JHU Johns Hopkins University, Baltimore
LC Lynchburg College, Lynchburg, Virginia
LHS Liberty High School and Bedford County Collection, Bedford, Virginia
LSUMZ Louisiana State University Museum of Zoology, Baton Rouge
MCZ Harvard University Museum of Comparative Zoology, Cambridge, Massachusetts
MLBS Mountain Lake Biological Station (University of Virginia), Pembroke, Virginia
MNHS Maryland Natural History Society, Baltimore
MSWB Museum of Southwestern Biology, University of New Mexico, Albuquerque
NCSM North Carolina State Museum of Natural History, Raleigh
NLU Northeast Louisiana University, Monroe
NSU Northwestern State University, Natchitoches, Louisiana
NVCC Northern Virginia Community College, Fairfax
ODU Old Dominion University, Norfolk, Virginia
OKG Private collection of O. K. Goodwin, Newport News, Virginia
OSU Museum of Zoology, Ohio State University, Columbus (B. D. Valentine Collection)
OU Stovall Museum, University of Oklahoma, Norman
PNSC Peninsula Nature and Science Center, Newport News, Virginia
RU Radford University, Radford, Virginia
RFC Private collection of R. F. Clarke, Emporia, Kansas (now part of Emporia State University Collection)
RH Private collection of Richard Highton, College Park, Maryland
RMWC Randolph-Macon Woman's College, Lynchburg, Virginia
SDSNH San Diego Society of Natural History, San Diego, California
SSM Savannah Science Museum
SNP Shenandoah National Park Collection, Luray, Virginia
TAMU Texas Cooperative Wildlife Collection, Texas A and M University, College Station
UAMNH University of Alabama, Museum of Natural History, Tuscaloosa
UC University of Colorado, Boulder
UFMNH University of Florida Museum of Natural History, Gainesville

UI	Museum of Zoology, University of Illinois, Urbana
UK	Museum of Natural History, University of Kansas, Lawrence (Fort A. P. Hill Collection)
UM	University of Minnesota, Minneapolis
UMMZ	University of Michigan Museum of Zoology, Ann Arbor (J. T. Wood Collection)
USFWS	U.S. Fish and Wildlife Service, Laurel, Maryland
USNM	U.S. National Museum, Smithsonian Institution, Washington, D.C.
VCU	Virginia Commonwealth University, Richmond
VIMS	Virginia Institute of Marine Science, Gloucester Point
VMI	Virginia Military Institute, Lexington
VMNH	Virginia Museum of Natural History, Martinsville
VPI	Virginia Polytechnic Institute and State University, Blacksburg (W. L. Burger Collection)
WCC	Wytheville Community College, Wytheville, Virginia
WLU	Washington and Lee University, Lexington, Virginia
WVBS	West Virginia Biological Survey, Marshall University, Huntington
YMP	Peabody Museum of Natural History, Yale University, New Haven

Questionable and Erroneous Records from Virginia

A confirmed specimen of the red-belly water snake (*Nerodia erythrogaster*) was taken in Fairfax County near Alexandria in 1974. This locality is approximately 100 miles north of the next nearest verified locality. Since no other specimens have been found, it must be assumed at this time that this specimen either escaped or was released. An earlier record of a red-belly water snake taken in the George Washington National Forest (Uhler, Cottam, and Clarke, 1939) also exists. This locality is even farther removed from the known range of this species. Since the specimen is no longer in existence, it must be assumed that its occurrence in this area was also the result of an escape or release.

The Field Museum of Natural History lists a southern hognose snake (*Heterodon simus*) from "Virginia." However, a thorough search of all specimens and skeletal material failed to locate the specimen for verification.

A snake identified as a "black kingsnake" (*Lampropeltis getulus niger*) was located in a display collection maintained by Virginia Military Institute. This snake had been collected in Lexington, Rockbridge County, in 1946. Following a preliminary examination, the specimen was taken to the National Museum of Natural History for determination. The specimen appears most similar to the speckled kingsnake (*Lampropeltis*

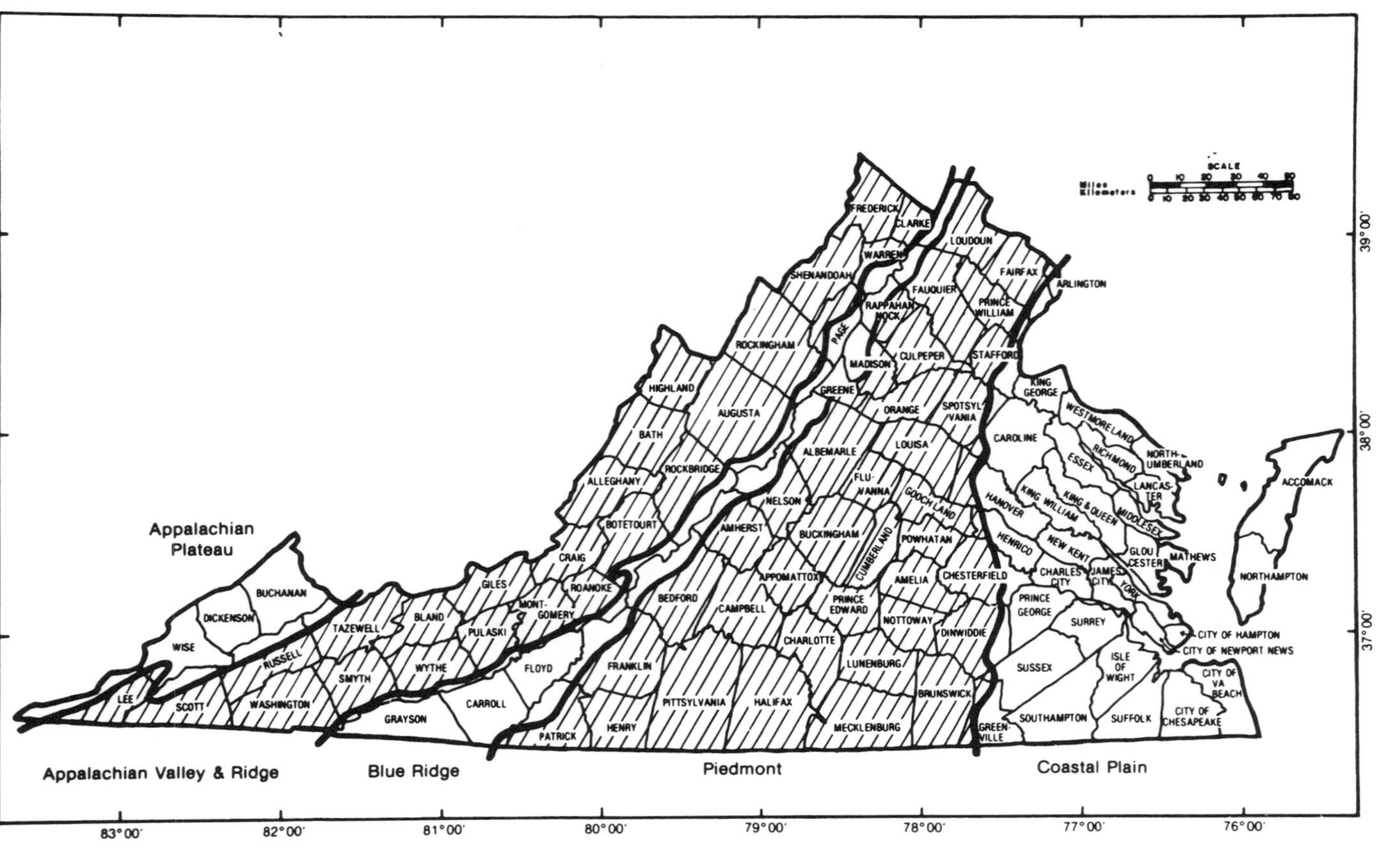

Fig. 7. Physiographic provinces of Virginia

getulus holbrooki), the nearest portion of whose range is in western Kentucky and western Tennessee, a distance of several hundred miles west of this locality. It was concluded, therefore, that this specimen probably either was released or escaped from captivity.

A cottonmouth (*Agkistrodon piscivorus*) in the National Museum of Natural History is recorded as having been taken in Arlington, Arlington County, in 1879. This locality is also approximately 100 miles north of the next nearest verified locality, thus making this record an unlikely natural occurrence. Unfortunately, this specimen could not be located for verification.

The reported occurrences in Virginia of two other species—the pygmy rattlesnake (*Sistrurus miliarius*) and the eastern diamondback rattlesnake (*Crotalus adamanteus*)—are discussed in the Pit Viper account (p. 123).

Physiographic Provinces of Virginia

The land surface of Virginia can be divided into five regions, or provinces, based on topographic features. From east to west these are: Coastal Plain, Piedmont Plateau, Blue Ridge, Ridge and Valley, and Appalachian Plateau (fig. 7).

The Coastal Plain extends westward from the Atlantic Ocean to the fall line, a belt of rapids or waterfalls along each stream as it flows from the relatively harder rocks of the Piedmont onto the relatively softer rocks of the Coastal Plain. The fall line thus acts as a barrier to tidal movements within Coastal Plain streams. Because these streams and rivers are affected by tides, the Coastal Plain region is often referred to as the "Tidewater" region of the state. Coastal Plain soils are composed largely of sands, gravels, clays, and marls. Elevation ranges from sea level to approximately 300 feet.

The Piedmont Plateau Province extends from the fall line on the east to the Blue Ridge Province on the west. For the most part it is a relatively low, rolling plateau region cut by many streams. The western portion, however, is distinctly more hilly than the eastern two-thirds of the province. Willis Mountain in Buckingham County is an exceptional topographic feature known in geological terminology as a monadnock. The Piedmont is basically well drained by such major rivers as the Potomac, Rappahannock, James, and Roanoke, but it does contain several distinctive lowland areas (Triassic Lowlands). Elevation ranges from approximately 300 feet to 1200 feet.

The Blue Ridge Province is a long, narrow, relatively rugged area bordered on the east by the Piedmont Plateau and on the west by the Ridge and Valley Province. It is part of the range that extends south to the Great Smokies where numerous mountain peaks rise above 6000 feet in elevation. Elevations in Virginia range from approximately 1200 feet up to the fir-capped peak of Mount Rogers in Grayson County which stands at 5729 feet. Mount Rogers is the highest point in Virginia. A large part of the Blue Ridge Province is covered by relatively thin mountain soils. Bedrock consists primarily of granites, slates, and greenstone. Rock exposures and talus slopes are locally common. The portion of this province northeast of the Roanoke River is drained by streams and rivers that flow into the Atlantic Ocean, while the portion of the province southwest of the Roanoke River is drained by watercourses that eventually lead into the Gulf of Mexico.

The Ridge and Valley Province, as its name implies, consists largely of fairly well defined valleys and intervening ridges. Like the Blue Ridge Province, the northern portion is drained by rivers flowing into the Atlantic Ocean and the southern portion is drained by streams and rivers that are part of the Gulf of Mexico drainage system. Two quite different topographic areas can be identified within this province—the Great Valley and the Alleghanies. The Great Valley (including the Shenandoah Valley) is a broad, gently undulating plain that lies immediately west of the Blue Ridge. This portion of the province varies in elevation from approximately 800 to 2500 feet. The Valley ranges in width from about 30 miles in northern Virginia to less than one mile near Buchanan in Botetourt County. Along the western edge of the Great Valley the topography changes abruptly. The Alleghanies are a series of alternating high narrow ridges and broader valleys with elevations at some points exceeding 4000 feet. The Ridge and Valley Province is underlain primarily by limestones, dolomites, sandstones, and shales. Shale barrens occur in several areas. Many springs, caves, and sinks occur in this region. Major rivers such as the Potomac, James, and Roanoke drain the northern part of the province and lead to the Atlantic Ocean, while the New and Tennessee river systems drain the southern portion and lead to the Gulf of Mexico.

The westernmost physiographic province in Virginia is known as the Appalachian Plateau. Only the seven southwesternmost counties are wholly or partly included in this province, which is the largest physiographic division in the Southern Appalachians. This province consists of an elevated plateau whose generally level to gently undulating surface has a gradual westward downslope. Elevations range from over 4100 feet to less than 900 feet along the Virginia-Kentucky state line. The bedrock consists mainly of sandstone, shale, limestone, and coal. Coal beds are abundant and have given rise to vast underground and strip mining operations. Most of the soils are fine sandy loams or silt loams. Drainage is primarily by means of the Big Sandy and Cumberland rivers which are part of the Ohio River basin.

Additional, more detailed, information concerning Virginia's physiographic provinces can be found in Fenneman (1938), Hoffman (1969), Dietrich (1970), and Woodward and Hoffman (1991).

KEYS TO THE SNAKES OF VIRGINIA

Two families of snakes occur in Virginia. They may be distinguished in the following manner:

1a. Pit present between the eye and nostril; hinged fangs present in anterior portion of mouth . . . . Family Viperidae (Subfamily Crotalinae)

1b. Pit absent; no hinged fangs in mouth . . . . Family Colubridae

KEY TO THE NONPOISONOUS SNAKES OF VIRGINIA (Family Colubridae)

1a. Keels on some or all of the dorsal scales 2
1b. All scales smooth 19
2a. Anal plate divided 3
2b. Anal plate undivided 17
3a. Either the preocular scale or the loreal scale missing; only one scale between the nasal scale and the eye 4
3b. Loreal and preocular scales present; two scales between the nasal scale and the eye 9
4a. Preocular scale present and not horizontally elongate; loreal scale absent 5
4b. Loreal scale present and horizontally elongate; preocular scale absent 6
5a. Scales in 15 rows . . . . *Storeria occipitomaculata*
5b. Scales in 17 rows . . . . *Storeria dekayi*
6a. Lower labial scales not more than 7; scale rows 15 or 17 7
6b. Lower labial scales 8 to 10; scale rows 19 8
7a. Scales strongly keeled; 1 postocular; 5 upper labials . . . . *Virginia striatula*
7b. Scales mostly smooth or nearly so; 2 postoculars; 6 upper labials . . . . *Virginia valeriae*
8a. Single internasal scale . . . . *Farancia abacura*

8b. 2 internasal scales . . . . *Farancia erytrogramma*
9a. Rostral scale turned upward and keeled . . . . *Heterodon platirhinos*
9b. Rostral scale not turned upward and not keeled 10
10a. Dorsal scales strongly keeled 11
10b. Dorsal scales weakly keeled 16
11a. Scale rows 17; bright green color in life (usually bluish-gray when preserved) . . . . *Opheodrys aestivus*
11b. Scale rows more than 17 12
12a. Scale rows 19 13
12b. Scale rows more than 19 14
13a. Light stripes present at sides of belly; belly with two parallel narrow dark brown stripes . . . . *Regina septemvittata*
13b. No light stripes present at sides of belly; belly with double row of distinct black spots or half-moons . . . .*Regina rigida*
14a. Scale rows usually 27 to 33; lower labials usually 11 to 13 . . . . *Nerodia taxispilota*
14b. Scale rows usually 21 to 25; lower labials usually 10 15
15a. Ventral surface marked with red and black half-moons . . . . *Nerodia sipedon*
15b. Ventral surface red or orangish-red and without markings . . . . *Nerodia erythrogaster*
16a. Bright reddish or brownish dorsal blotches outlined in black; dorsal markings form spear point between the eyes . . . . *Elaphe guttata*
16b. Shiny black above . . . . *Elaphe obsoleta*
17a. 19 scale rows; 2 prefrontals; rostral scale does not penetrate between internasal scales 18
17b. More than 19 scale rows; usually 4 prefrontals; rostral scale penetrates between internasal scales . . . . *Pituophis melanoleucus*
18a. Lateral stripe involving 2d and 3d scale rows above belly . . . . *Thamnophis sirtalis*
18b. Lateral stripe involving 3d and 4th scale rows above belly . . . . *Thamnophis sauritus*
19a. Anal plate divided 20
19b. Anal plate undivided 27
20a. Either the preocular scale or the loreal scale missing; only one scale between the nasal scale and the eye 21
20b. Loreal scale and preocular scale present; two scales between the nasal scale and the eye 25

21a. Preocular scale present and not horizontally elongate; loreal absent . . . . *Tantilla coronata*
21b. Loreal scale present and horizontally elongate; preocular absent 22
22a. 13 scale rows . . . . *Carphophis amoenus*
22b. More than 13 scale rows 23
23a. Lower labials not more than 7; scale rows 15 or 17 . . . . *Virginia valeriae*
23b. Lower labials 8 to 10; scale rows usually 19 24
24a. Single internasal scale . . . . *Farancia abacura*
24b. 2 internasal scales . . . . *Farancia erytrogramma*
25a. Lower preocular small and wedged between two of the upper labials . . . . *Coluber constrictor*
25b. Lower preocular of normal size and position 26
26a. Gray to bluish-black above with yellowish to reddish neck ring (neck ring may be incomplete) . . . . *Diadophis punctatus*
26b. Bright green above (usually bluish-gray when preserved) . . . . *Opheodrys vernalis*
27a. Snout pointed; rostral scale projects beyond lower jaw; ventral surface light without markings . . . . *Cemophora coccinea*
27b. Snout normal; rostral scale smaller; ventral surface with at least some dark markings 28
28a. Pattern without red, and without rings or black-bordered blotches . . . . *Lampropeltis getula*
28b. Pattern with red, or with rings or black-bordered blotches 29
29a. Dorsum shiny and iridescent with 3 red stripes on a bluish-black background . . . *Farancia erytrogramma*
29b. Dorsum not as above 30
30a. Pattern of rings, or if of blotches, then these broadly in contact with 5th or lower row of scales . . . . *Lampropeltis triangulum*
30b. Pattern of black-edged dorsal blotches extending no lower than upper edge of 5th row of scales . . . . *Lampropeltis calligaster*

KEY TO THE POISONOUS SNAKES OF VIRGINIA (Family Viperidae, Subfamily Crotalinae)

1a. Rattle present 2
1b. Rattle absent 3
2a. Yellowish to dark brown ground color; dark brown or black V-shaped crossbands; no

middorsal stripe; no band between eye and angle of jaw . . . . *Crotalus horridus horridus*

2b. Pale grayish-brown to flesh pink ground color; dark brown or black V-shaped crossbands; mid-dorsal reddish-brown stripe; dark oblique band runs posteriorly from eye to beyond angle of jaw . . . . *Crotalus horridus atricaudatus*

3a. 25 scale rows. Loreal scale absent . . . . *Agkistrodon piscivorus*

3b. 23 scale rows. Loreal scale present . . . . *Agkistrodon contortrix*

COMPARISON TABLES TO AID IN IDENTIFICATION

These tables are designed for quick comparison to assist with identification of species with certain similar traits. They are referred to throughout the species accounts under the "Similar Species" heading.

Table A. Adult Semiaquatic Snakes

Species	Anal plate	Scales	Dorsal pattern	Ventral pattern	Other distinguishing characteristics
Northern Water Snake	Divided	Keeled 23–25 rows	Reddish to dark brown saddles and blotches on gray or brown; may be uniform brown	Red or black half-moons	No facial pit; double row of scales under tail
Red-Belly Water Snake	Divided	Keeled 23–25 rows	Uniformly reddish-brown	Unmarked red or orange-red	No facial pit; double row of scales under tail
Brown Water Snake	Divided	Keeled 27–33 rows	Square brown blotches on lighter brown; often indistinct	Dark half-moons	No facial pit; double row of scales under tail
Cottonmouth	Undivided	Weakly keeled 25 rows	Dark crossbands; often indistinct	Irregular dark stippling and blotches	Facial pit; single row of scales under tail

Table B. Juvenile Semiaquatic Snakes

Species	Anal plate	Scales	Dorsal pattern	Ventral pattern	Other distinguishing characteristics
Northern Water Snake	Divided	Keeled 23–25 rows	Dark brown or black saddles on gray background	Dark half-moons	No facial pit; double row of scales under tail
Red-Belly Water Snake	Divided	Keeled 23–25 rows	Dark brown blotches on orange-brown background	Uniformly red or orange-red	No facial pit; double row of scales under tail
Brown Water Snake	Divided	Keeled 27–33 rows	Square brown blotches on lighter brown background	Dark half-moons	No facial pit; double row of scales under tail
Cottonmouth	Undivided	Weakly keeled 25 rows	Reddish-brown crossbands on lighter brown background; yellow tail tip; broad dark band through eye	Brown blotches	Facial pit present; single row of scales under tail

Table C. "Black" Snakes

Species	Anal plate	Scales	Dorsal pattern	Ventral pattern	Other distinguishing characteristics
Black Rat Snake	Divided	Weakly keeled 25–27 rows	Shiny black, often with scattered light specks	Yellowish-white with scattered black squares; grayish toward rear	Cross section of body shaped like loaf of bread
Black Racer	Divided	Smooth 17 rows	Dull black	White chin; remainder bluish-gray	Body round in cross section
Eastern Kingsnake	Undivided	Smooth 21 rows	Shiny black with distinct white chain pattern	White or yellow squares on bluish-black or vice versa	Head not distinctly wider than neck
Black King Snake	Undivided	Smooth 21 rows	Shiny black with light specks or indistinct pattern as above	White or yellow squares on bluish-black or vice versa	Found only in extreme southwestern Virginia
Eastern Hognose Snake	Divided	Keeled 23–25 rows	Variable: uniform black or gray or with pattern (see Table D)	Usually gray with darker mottling	Has upturned snout; usually hisses, flattens neck, and may "play dead"

Species	Anal plate	Scales	Dorsal pattern	Ventral pattern	Other distinguishing characteristics
Copperhead	Undivided	Weakly keeled 23 rows	Brown hourglass-shaped crossbands on lighter brown background	Yellowish-brown with dark spots on sides	Facial pit; copper-colored head; single row of scales under tail
Eastern Hognose Snake	Divided	Keeled 23–25 rows	Variable: generally dark blotches and/or spots on yellow, orange, brown, or gray; may be solid black or gray (see Table C)	Indistinct gray blotches on greenish-yellow; may be plain dark gray	Upturned snout; usually hisses, flattens neck and may "play dead"
Corn Snake	Divided	Weakly keeled 27 rows	Bright red blotches ringed with black on orange or gray background	White with squarish black blotches	Stripes under tail; red spear point between eyes
Eastern Milk Snake	Undivided	Smooth 19 rows	Brown or reddish-brown blotches bordered in black on gray or tan background	Checkerboard pattern of black and white	V- or Y-shaped marking on neck; head barely distinct
"Coastal Plain Milk Snake" (intergrade between eastern milk snake and scarlet kingsnake)	Undivided	Smooth 19 rows	Large reddish blotches bordered with black on white or yellow background	White or yellow with black squares	Light collar on neck; head barely distinct

Table D. Blotched Snakes

Species	Anal plate	Scales	Dorsal pattern	Ventral pattern	Other distinguishing characteristics
Mole Kingsnake	Undivided	Smooth 19–23 rows	Variable: uniform olive-brown or brown with well-separated reddish-brown blotches	Yellowish-brown with indistinct brown spots	Head barely distinct
Northern Pine Snake	Undivided	Keeled 29 rows	Black and dark brown blotches on white or gray background	White with dark spots on sides	Hisses loudly; spotty distribution
Black Rat Snake (Juvenile)	Divided	Weakly keeled 25–27 rows	Dark gray or brown squarish blotches on light gray background	Yellowish or white with black squares	Dark stripe through eyes; body bread loaf-shaped in cross section
Black Racer (Juvenile)	Divided	Smooth 17 rows	Oval dark gray blotches on gray or bluish-gray background	Gray with dark dots	Body round in cross section; tail gray

Table E. Snakes with Neck Patterns

Species	Anal plate	Scales	Dorsal pattern	Ventral pattern	Neck
Northern Ringneck Snake	Divided	Smooth 15 rows	Slate to bluish-black	Yellowish-orange, sometimes with rows of dark spots	Narrow yellow ring. Found throughout Virginia except southeast portion and Eastern Shore
Southern Ringneck Snake	Divided	Smooth 15 rows	Bluish-black	Orange or red with row of dark spots down center	Orange ring, often incomplete. Southeastern Virginia and Eastern Shore
Southeastern Crowned Snake	Divided	Smooth 15 rows	Brown or reddish-brown	Pinkish- or yellowish-white; no spots	Light band across back of head followed by black collar.
Northern Brown Snake (Juvenile)	Divided	Keeled 17 rows	Grayish-brown	Yellowish-brown with dark spots on sides	Yellow collar
Red-Belly Snake	Divided	Keeled 15 rows	Gray to brown, usually with four narrow dark stripes	Bright red or orange-red	2 or 3 light spots

Table F. Small "Brown" Snakes

Species	Anal plate	Scales	Dorsal pattern	Ventral pattern	Other distinguishing characteristics
Northern Brown Snake	Divided	Keeled 17 rows	Brown with 2 rows of black dots down back	Yellowish-brown with dark dots on sides	Dark streak on side of head
Rough Earth Snake	Divided	Keeled 17 rows	Gray to red reddish-brown	Cream, white, or greenish-white	Pointed snout; five upper labials
Smooth Earth Snake	Divided	Smooth or weakly keeled 15–17 rows	Gray to brown, with scattered black dots	Cream; may have pinkish tinge	Rounded snout; six upper labials
Eastern Worm Snake	Divided	Smooth 13 rows	Brown; lower sides pinkish	Pink	Spinelike tail tip
Red-Belly Snake	Divided	Keeled 15 rows	Gray to brown, usually with 4 narrow dark stripes	Bright red or orange-red	2 or 3 light spots on neck

Species Accounts

HARMLESS SNAKES
Family Colubridae

The family Colubridae is the largest family of snakes in Virginia as well as in the United States and the world. Approximately 75 percent of all North American snakes are included in this family. Twenty-seven of the 30 species of snakes found in Virginia are colubrids. They range in size from the tiny worm and earth snakes to the large pine and rat snakes which may attain lengths of over six feet. The colubrids occupy a wide variety of habitats with some being aquatic, some terrestrial, some arboreal, and a few fossorial.

WATER SNAKES
(Genus *Nerodia*)

Three members of the genus *Nerodia*, or water snakes, inhabit Virginia. These snakes are aptly named because water is the common ingredient in the various habitats where they occur. From rippling mountain brooks to lowland dark-water swamps, these semiaquatic reptiles are often the most conspicuous element in local snake populations. Members of this group frequently rest in branches that overhang the water and drop into the water at the slightest disturbance. All of these snakes are good swimmers.

All members of the genus *Nerodia* have strongly keeled dorsal scales and a divided anal plate. Scale rows range between 23 and 33.

Water snakes feed primarily upon frogs, toads, salamanders, fish, and crayfish. Other food items include worms, crustaceans, and insects. Contrary to popular belief, water snakes are not a threat to local fish populations. Under natural conditions these snakes capture few game fish. Usually it is the slower moving species that are eaten, as well as diseased or even dead individuals. In fact, studies have shown that heavy predation can actually benefit fishing by reducing or preventing an overpopulation of stunted panfish.

All members of the genus *Nerodia* give birth to living young rather than laying eggs. Mating generally takes place in the spring with the young normally being born between August and early October. Litters may consist of 50 or more individuals, although most litters contain between 20 and 40 snakes. Newborn snakes range between 8 and 12 inches in total length and are normally strongly patterned.

Water snakes are well known for their aggressive behavior. Most will flatten their bodies, strike, and bite when frightened. When first handled, many will secrete a foul-smelling substance from glands near the base of their tail.

Some of the water snakes are very often mistaken for the poisonous cottonmouth, or "water moccasin." Although reports of cottonmouths come from all parts of Virginia, the range of the true cottonmouth covers only the southeastern portion of the state. From a distance, however, adult individuals of each of the water snakes can look very similar. Upon closer inspection several features can be used to distinguish between these snakes. The poisonous cottonmouth possesses (1) a sensory pit between each eye and nostril; (2) a pair of fangs projecting downward in the anterior portion of the mouth; (3) elliptical pupils in their eyes; and (4) a single row of scales on the underside of the tail. In addition, the cottonmouth will almost always "stand its ground" when disturbed, holding its mouth open in readiness to strike. This posture serves to reveal not only the fangs but also the cottony-white lining of the mouth which has given this snake its common name of cottonmouth. The nonpoisonous water snakes (1) lack sensory pits; (2) have no fangs; (3) have round pupils in their eyes; and (4) possess a double row of scales on the underside of the tail. Furthermore, these snakes generally do not "stand their ground" when alarmed and tend to retreat more easily. The experienced observer can also distinguish a difference in the structure of the head. The top of the head of the cottonmouth is very flat and meets the side of the face at a very sharp angle. In nonpoisonous snakes the head and facial region is usually much more rounded. If in doubt, however, a person unfamiliar with snakes and unable to make a positive identification should leave the snake alone.

Some authorities recognize *Natrix* as the sole genus of nonpoisonous water snakes in the United States (Conant, 1975; et al.). Others recognize two genera: *Natrix* with six or seven species and *Regina* with four species (Smith and Huheey, 1960; Rossman, 1963b). When applied to Virginia, the latter classification places the brown, red-belly, and northern water snakes in the *Natrix* subgroup and the queen snake and glossy crayfish snake in the *Regina* subgroup. Investigations of blood proteins, chromosomes, scutellation, cranial bone structure, and hemipenes by Rossman and Eberle (1977) have resulted in the revival of the name *Nerodia* for members of the genus *Natrix* in North America. The genus *Regina* has been retained for the four species formerly assigned to it. We have elected to follow this most recent revision in this book.

Brown Water Snake
(*Nerodia taxispilota*)

Plate 1

Other Common Names: Water moccasin, water pilot, water rattle.

Description. Adult: Large, square, dark brown blotches on a lighter brown background. One row of squares runs down the back with an alternating row on each side. Belly is yellowish-brown with dark crescent-shaped or trapezoidal blotches. Pattern darkens and becomes less distinct with age. Snakes basking in the sun are often a plain dusky brown, but pattern reappears when wet.
Juvenile: Similar to adult but lighter.
Scalation: Dorsal scales strongly keeled; 27–33 scale rows; anal plate divided. Loreal and preocular scales present (fig. 8).

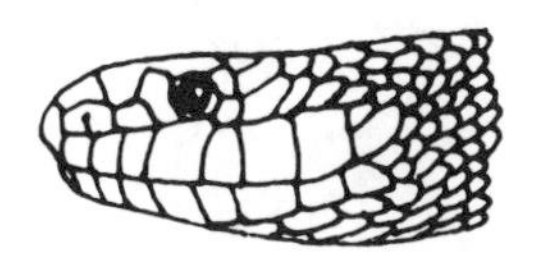

Fig. 8. Head of brown water snake, natural size

Size: Approximately 8 inches at birth to over 60 inches in some adults. One snake from Pungo Township, Virginia, measured 69 1/2 inches (Werler and McCallion, 1951). Most adults are 36 to 48 inches in length. These snakes are heavy-bodied with a 48-inch specimen weighing as much as 3 1/3 pounds.
Variation: No subspecific variation is recognized.
Similar Species: Other nonpoisonous water snakes and the cottonmouth. See Tables A and B above.

Habitat. Although found in a variety of quiet waters, the brown water snake tends to be more common than other water snakes in larger bodies of water. These snakes are often seen basking in the crotches of cypress trees growing several hundred feet from shore in Lake Drummond in the Great Dismal Swamp. Brown water snakes are common along many of the other lakes, canals, and rivers in southeastern Virginia. Neill (1958) has recorded this species in association with brackish water.

Range. This snake ranges from southeastern Virginia to Florida and west to eastern Alabama. Its range in Virginia is restricted to the southeastern portion of the state.

Habits. During the day, brown water snakes spend much time basking in the sun, possibly more so than other water snakes. They may climb to a height of 15 or 20 feet above the water and lay for hours without stirring. A quiet approach at this napping time can mean success for the collector if the specimen is not too high to reach. This species tends to be

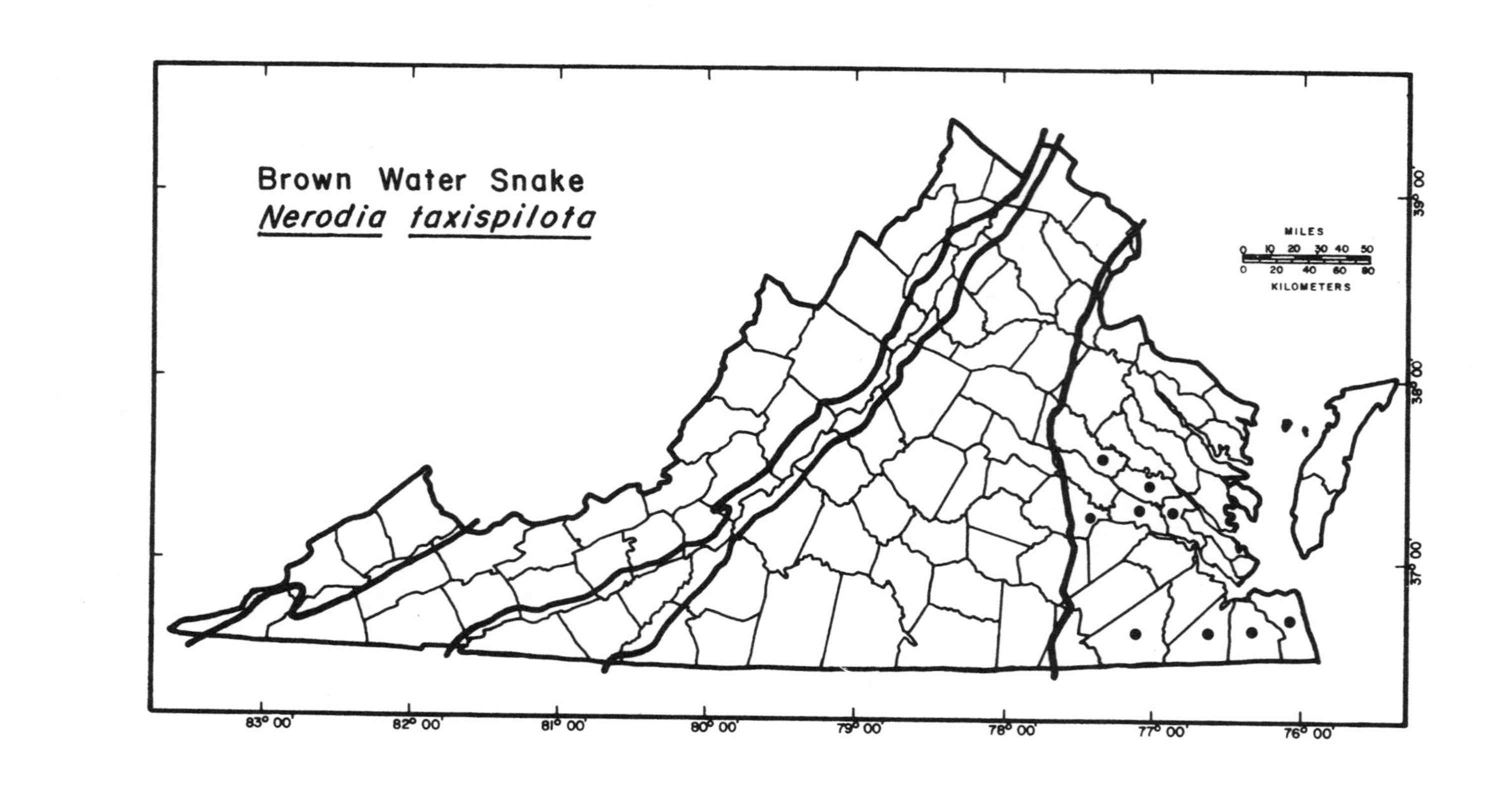
Brown Water Snake
Nerodia taxispilota
MILES
0 10 20 30 40 50
0 20 40 60 80
KILOMETERS
39° 00'
38° 00'
37° 00'
83° 00'
82° 00'
81° 00'
80° 00'
79° 00'
78° 00'
77° 00'
76° 00'

less vicious than other water snakes and sometimes may not attempt to bite when captured. However, they usually will thrash and twist in efforts to escape.

Reproduction. Females bear living young. Between 12 and 50 young are normally born in late summer. Newborn snakes range in length from 8 to 14 inches.

Food. Fish comprise the primary food of these snakes, although frogs and other aquatic animals are also eaten. In captivity, brown water snakes will eat whole bluegills and other sunfish. As with other water snakes, the prey is swallowed alive.

Enemies. Large individuals, because of their impressive size, have relatively few enemies other than man, who too often shoots them with the belief that they are dangerous "water moccasins." Juvenile and small individuals are preyed upon by raccoons, mink, herons, kingsnakes, and other shoreline predators.

In Captivity. Brown water snakes generally do well in captivity. They can be kept in the same type of cage as more terrestrial species and do not require water in which to submerge. Because of their size, it is often practical to keep adults in an outdoor enclosure during the warmer months of the year.

Folklore. Two bits of folklore are associated with this species. One is that it is a deadly type of aquatic rattlesnake (hence the name "water rattle"), although it lacks any sort of rattle. Brown water snakes are also called "water pilots" in the belief that they guide other snakes to safety.

Location of Specimens. AMNH, ANSP, CM, GMW, LSUMZ, NVCC, UMMZ, USFWS, USNM, VCU, VPI and SU.

Red-belly Water Snake
(*Nerodia erythrogaster*)

Other Common Names: Copperbelly, moccasin. *Plate 2*

Description. Adult: Solid brown, reddish-brown, or dark olive above with a red or orangish-red belly. No blotches, spots, or other markings are usually evident on adults.
Juvenile: Brown blotches down the back with smaller blotches on the sides. Orange-brown ground color. Belly is red or orange-red and without markings.
Scalation: Dorsal scales strongly keeled; 23–25 scale rows; anal plate divided. Loreal and preocular scales present (fig. 9).
Size: Approximately 10 inches at birth to over 5 feet in some adults. Adults are heavy-bodied and are usually between 30 and 48 inches in length.

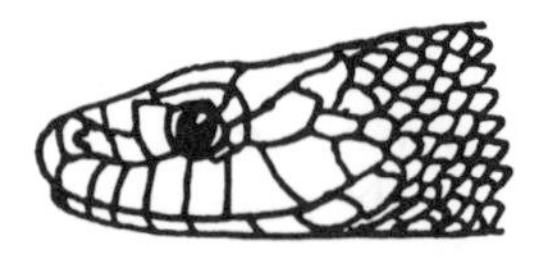

Fig. 9. Head of red-belly water snake, natural size

Variation: *Nerodia erythrogaster erythrogaster* (Holbrook), the red-belly water snake, is the only subspecies occurring in the state.
Similar Species: From a distance they might be confused with other water snakes and the cottonmouth (see Table A above), but upon closer inspection the bright ventral color and the lack of markings, both above and below, on the adults should distinguish them. For juveniles, see Table B above.

Habitat. River swamps choked with semiaquatic vegetation and characterized by bald cypress, red maple, and sweet gum provide the favored habitat. Red-belly water snakes are also often very abundant in the smaller acid waters within their range. These are the most numerous snakes of the dark cypress and gum pools in the Seashore State Park Natural Area. These pools are separated by sandy ridges of pine and hardwood into which red-belly water snakes often wander during the summer. These two habitats provide a favorable combination which supports a large population. Red-belly water snakes are common in certain areas of the Great Dismal Swamp and in some of the dark-water rivers of southeastern Virginia.

Range. The range of this species extends from southern Delaware and southern Maryland to northern Florida and eastern Alabama. In Virginia, it is found in the southeastern portion of the state. Musick (pers. comm.) noted that an isolated population exists in a swamp within the Newport News City Park. A confirmed red-belly water snake was taken near Alexandria, Fairfax County, in 1974 (Boo, 1974; *Va. Herp. Soc. Bull.* no. 76, Jan.–Feb. 1975). Unless a colony of these snakes can be found inhabiting the area, this single specimen must be considered as either having escaped from captivity or having been liberated near where it was found.

Habits. This species is the most terrestrial of the three species of *Nerodia* inhabiting Virginia and will wander hundreds of yards into relatively dry forests, particularly during hot summer weather. When resting near water, these snakes tend to select a spot directly on the ground rather than a spot high on a tree branch (as is often the case with the brown water snake) or on a log or in a low clump of vegetation (as is often the case with the northern water snake).

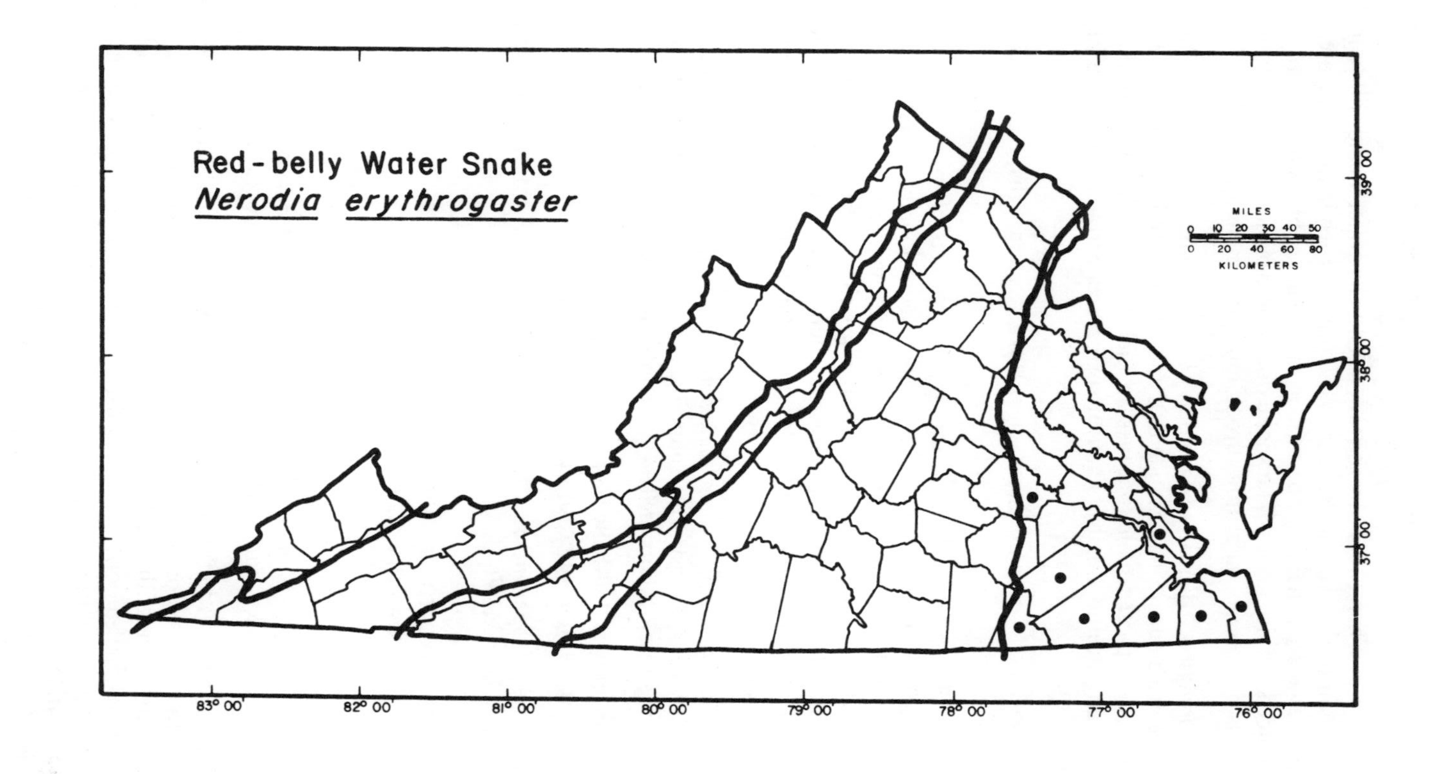
Red-belly Water Snake
Nerodia erythrogaster
MILES
0 10 20 30 40 50
0 20 40 60 80
KILOMETERS
39° 00'
38° 00'
37° 00'
83° 00'
82° 00'
81° 00'
80° 00'
79° 00'
78° 00'
77° 00'
76° 00'

When caught by the tail, a red-belly water snake will twirl off a portion of the tail and leave it to the predator while the snake escapes. This defensive reaction is common to many snakes but is particularly pronounced in this species. The snake does not grow a new tail as do many lizards but carries the stump throughout the remainder of its life.

Reproduction. Females bear up to 50 living young. Newborn young are between 10 and 12 inches in length and are born in late summer or early fall. One typical female that was 42 inches in total length gave birth to 29 young in mid-August. Each offspring was a fraction over 10 inches in length.

Longevity. Maximum known age: 8 years, 10 months, 2 days (Bowler, 1977).

Food. The primary food of this species consists of frogs and fish, but crayfish and salamanders are also taken. Occasionally, observers of animal life in the swamps at Seashore State Park will notice a commotion in the shallows of a cypress pond. It appears that a water snake is feeding on small fish it has cornered. However, in most cases a closer look reveals that the splashing and swirling of the water is caused by a male bowfin fish herding its young.

Enemies. Red-belly water snakes are taken by such predators as raccoons, herons, kingsnakes, snapping turtles, and large fish. Because of their larger size, adults are less susceptible to most predators, except man.

In Captivity. Red-belly water snakes do well in captivity. A dry cage with just enough water for drinking is suitable.

Location of Specimens. AMNH, CHE, CM, CWM, GMW, NVCC, USFWS, USNM, VCU, VIMS, VPI and SU.

Northern Water Snake
(*Nerodia sipedon*)

Plates 3 and 4

Other Common Names: Water moccasin, banded water snake, common water snake, midland water snake, dry-land moccasin.

Description. Adult: Reddish to dark brown crossbands and blotches on a lighter gray or brown background. Pattern often fades with age with old adults being almost black or dark brown. Belly is marked with red and black half-moons. An amelanistic specimen with a mostly white venter and pinkish brown eyes was recorded from Botetourt County (Robertson, 1985).
Juvenile: Dark brown or black crossbands and blotches on a gray background.

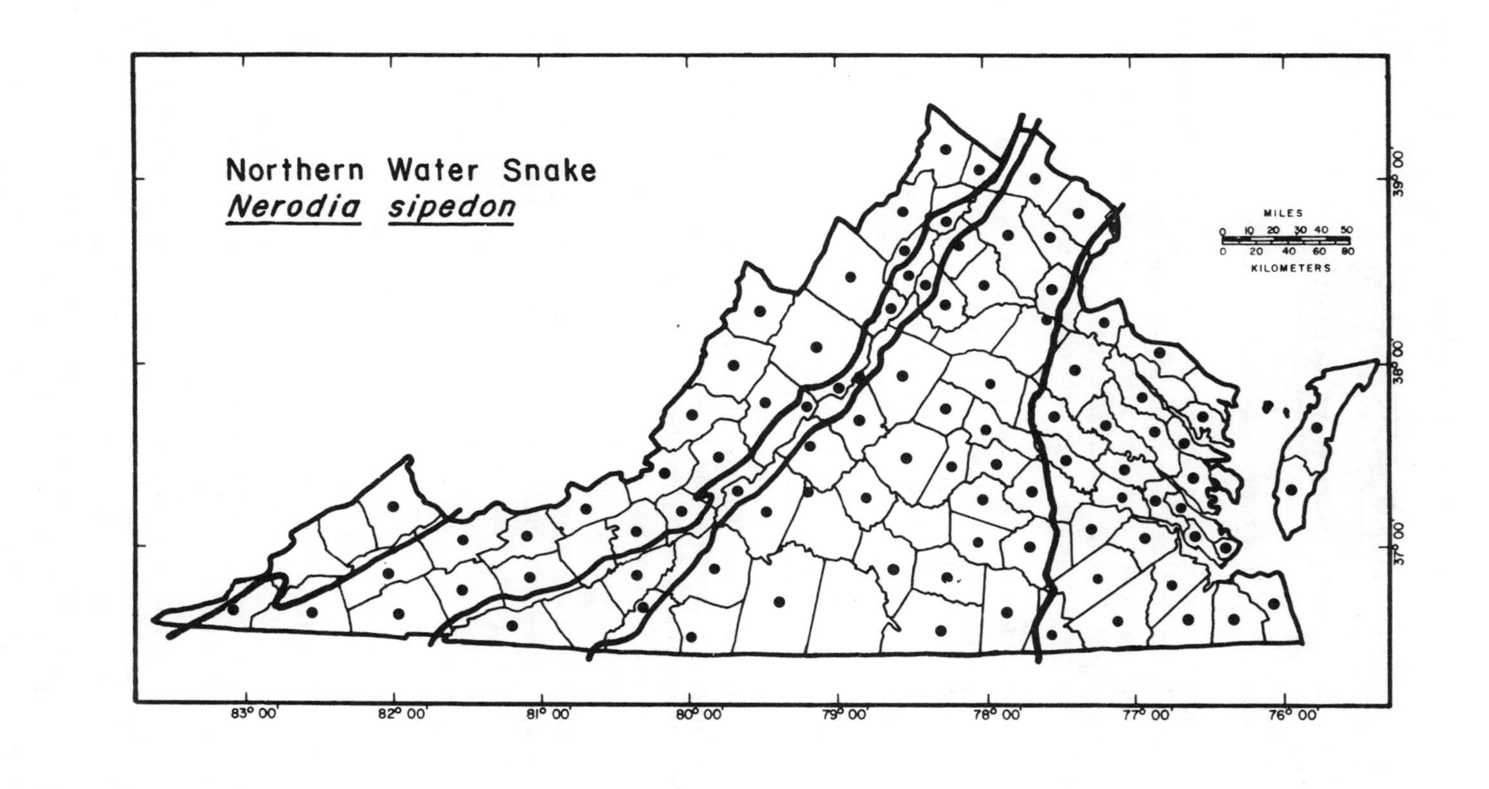
Northern Water Snake
Nerodia sipedon
MILES
0 10 20 30 40 50
0 20 40 60 80
KILOMETERS
39° 00'
38° 00'
37° 00'
83° 00'
82° 00'
81° 00'
80° 00'
79° 00'
78° 00'
77° 00'
76° 00'

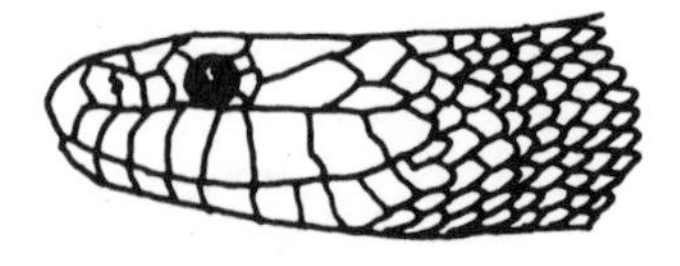

Fig. 10. Head of northern water snake, natural size

Scalation: Dorsal scales strongly keeled; 23–25 scale rows; anal plate divided. Loreal and preocular scales present (fig. 10).
Size: Approximately 8 inches at birth to over 48 inches in some adults. Most adults are between 24 and 36 inches.
Variation: Two subspecies are recognized in Virginia.

Plate 3

Nerodia sipedon sipedon (Linnaeus). Northern Water Snake. Distinct crossbands on anterior portion of body, but with alternating blotches on posterior portion. Dark markings on back and sides are wider than the spaces between them. Belly variable ranging from distinct half-moon-shaped markings to dusky mottling. Occurs statewide.

Plate 4

Nerodia sipedon pleuralis Cope. Midland Water Snake. Distinct crossbands on anterior portion of body, but with alternating blotches on posterior portion. Dark markings on back and sides are smaller than the spaces between them. Belly usually with two rows of half-moon-shaped markings. Reported only from Lee County (Burger, 1975).
Similar Species: Likely to be confused with other semiaquatic species (see Tables A and B above) and possibly the copperhead, hognose snake, and certain other snakes (see Table D above).

Habitat. Northern water snakes occur in a great variety of freshwater habitats ranging from small woodland streams to the shorelines of large rivers and reservoirs. Farm ponds with well-vegetated shorelines provide optimal habitat, sometimes containing unbelievable numbers of these snakes. In the southeastern portion of Virginia the common water snake tends to be found in smaller bodies of water than the brown water snake but in larger bodies of water than the red-belly water snake. Neill (1958) noted the occurrence of this species in brackish water.

Range. Northern water snakes range from Maine and southern Canada south to North Carolina, Georgia, and Alabama. The range extends westward to Minnesota, Nebraska, and Colorado. This species occurs throughout Virginia.

Habits. The northern water snake is usually very active at night. Large numbers can often be observed in a pond several feet from shore with their heads protruding above the water. Others roam the cattails looking for frogs and small fish. During the day, these water snakes may be observed sunning themselves on a low branch or log, resting in some hidden retreat, or perhaps continuing the search for food. Homing has been reported by Fraker (1970).

Reproduction. Females bear living young. Eight to 40 or more young are born during late summer and early fall. Dunn (1915*d*) recorded 40 young born on October 12 to a female taken in Nelson County.

Longevity. Maximum known age: 7 years, 4 months, 7 days (Bowler, 1977).

Food. Depending upon the season and the specific habitat of the individual snake, the primary foods are usually frogs, fish, crayfish, salamanders, or a combination of these. Other prey may include larval amphibians and insects. Uhler, Cottam, and Clarke (1939) examined 30 stomachs of this species taken from the George Washington National Forest in Virginia. Major food items, by volume, included nongame fish 48%, frogs 19%, game and pan fish 13%, salamanders 13%, and toads 3%.

Enemies. Northern water snakes are eaten by many of the predators that roam the shorelines and waters including raccoons, mink, otters, herons, kingsnakes, and snapping turtles. Young snakes are more susceptible to predation than adults and may also be preyed upon by bullfrogs and bass. Large water snakes can defend themselves well against many enemies. One of us (Clifford) once found a large specimen recuperating from a battle with a raccoon, according to tracks. The snake lost several inches of its tail and sustained some gashes, but it was surviving. The coon also survived, no doubt! In many situations, man is a substantial enemy of northern water snakes.

In Captivity. The northern water snake does well in captivity but, because of its vicious temperament, it makes a less than ideal pet. Feeding is not a problem, especially if a supply of goldfish or minnows is available. In fact, some captives will eat just about any animal matter that smells of fish.

Folklore. The superstition most often connected with this species is that it is poisonous. The "venomous water moccasins" reported from mountain, piedmont, and northern coastal plain counties of Virginia are actually harmless northern water snakes (as are many of those killed even within the cottonmouth's range).

Location of Specimens. AMNH, ANSP, CFR, CM, CU, CWM, DS, DU, ESU, FMNH, GMU, GMW, INHS, JHU, LC, MCZ, MLBS, NVCC, ODU, OSU, USFWS, UMMZ, UK, USNM, VCU, VIMS, VMNH, VPI and SU.

Crayfish Snakes (Genus *Regina*)

Two members of the genus *Regina* are found in Virginia—the queen snake and the glossy crayfish snake. Both of these species have strongly keeled dorsal scales and a divided anal plate. Each of these snakes possesses 19 scale rows.

These snakes are among the most aquatic and secretive of Virginia's snakes. They feed largely upon crayfish, small fish, and salamanders.

Queen Snake
(*Regina septemvittata*)

Plates 5 and 6

Other Common Names: Garter snake, willow snake, branch snake, leather snake, water snake.

Description. Adult: Olive brown to dark brown above with three indistinct narrow dark stripes. A broader yellow stripe involving the first and second scale rows runs along each side of the body. Underneath, two parallel narrow dark brown stripes run down the cream or yellow belly, each flanked by a broader stripe to the outside. These stripes merge toward the tail in adult snakes. The head is small.
Juvenile: Similar to adult.
Scalation: Dorsal scales strongly keeled; 19 scale rows; anal plate divided. Loreal and preocular scales present (fig. 11).

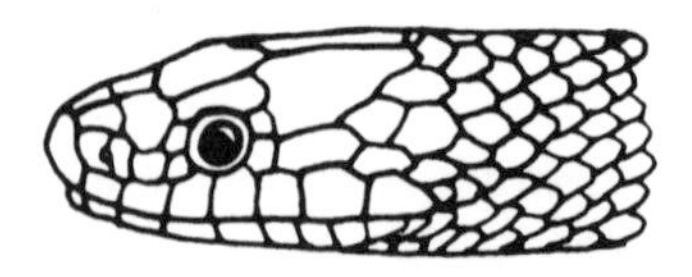

Fig. 11. Head of queen snake, 2× natural size

Size: Approximately 8 inches at birth to 36 inches. Most adults are between 18 and 24 inches long. Queen snakes are much more slender than most other water snakes.
Variation: No subspecific variation is recognized.
Similar Species: The glossy crayfish snake has two rows of black spots down the belly, a larger head, a stouter body, shiny scales, and a restricted and separate range. The eastern garter snake has a single anal plate and (usually) a light stripe down the center of the back.

Habitat. Queen snakes are most common along small, rocky creeks in hilly or mountainous areas. These streams usually consist of swift-flowing stretches alternating with quiet pools. They typically are lined with alders and willows and contain many large rocks and stones that provide hiding places for the snakes. Queen snakes may sometimes be found along rivers or quiet meadow streams.

Range. Queen snakes are found from southern Canada, New York, Michigan, and Wisconsin southward to northern Florida and eastern Mississippi. An isolated population exists in Arkansas and Missouri. In

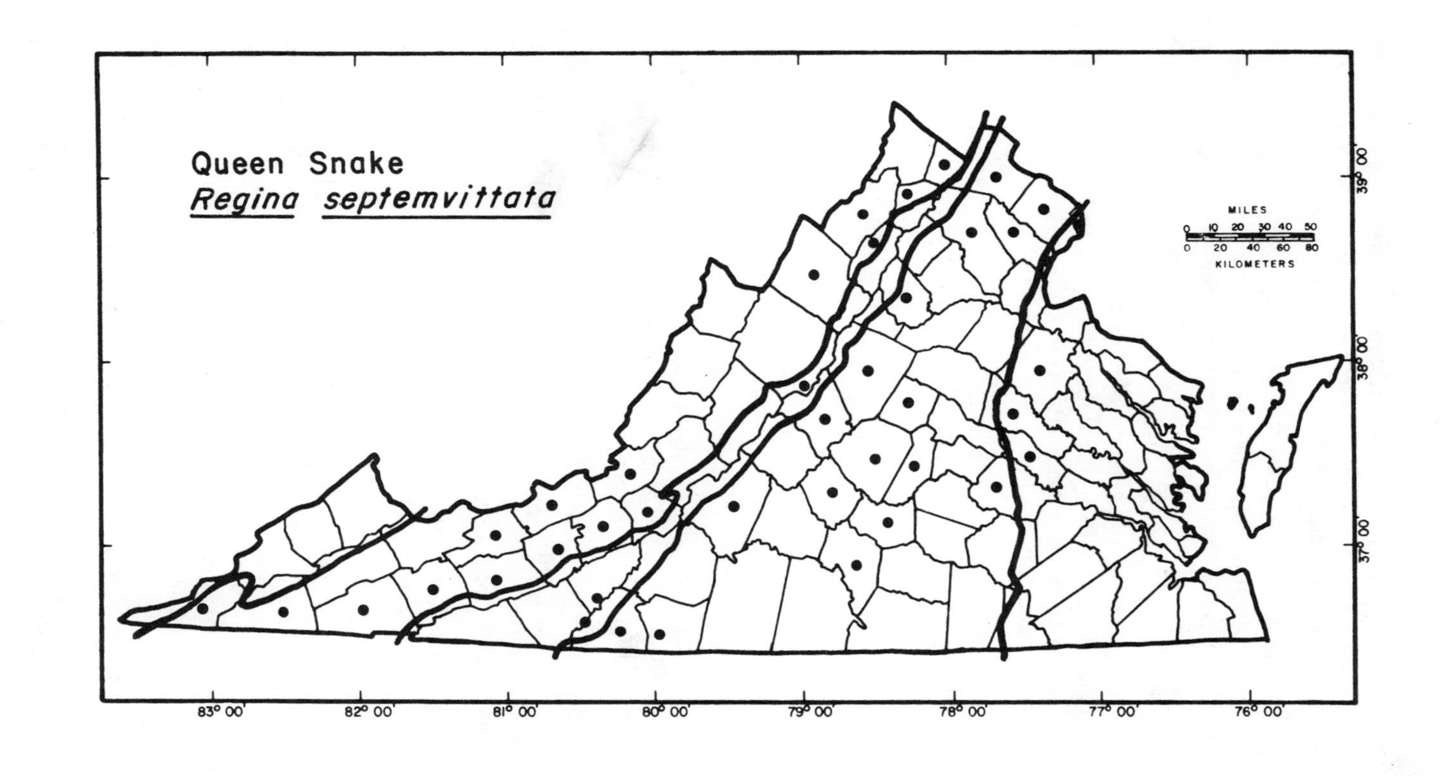
Queen Snake
Regina septemvittata
MILES
0 10 20 30 40 50
0 20 40 60 80
KILOMETERS
39° 00'
38° 00'
37° 00'
83° 00'
82° 00'
81° 00'
80° 00'
79° 00'
78° 00'
77° 00'
76° 00'

Virginia, this species occurs primarily in the northern and western portions of the state. It has been recorded as far east as Caroline, Hanover, and Chesterfield counties.

Habits. These snakes are very aquatic and are speedy swimmers in shallow waters. Much of their time is spent in the water, either resting or prowling. Queen snakes also climb into the vegetation overhanging streams and drop into the water when startled. At other times these secretive snakes hide under rocks in the water or along the bank. Queen snakes are among the most cold-tolerant of Virginia's snakes. Not only do they live in cold creeks, but they enter hibernation late and emerge early. There are many reports of large aggregations of these snakes gathering in small areas before hibernation.

Reproduction. Females bear living young. Up to 18 young, each between 7 and 9 inches in length, are normally born in August and September.

Longevity. Maximum known age: 19 years, 3 months, 17 days (Bowler, 1977).

Food. The principal food of the queen snake is crayfish in the soft-shell stage. Individuals may also occasionally eat small hard-shell crayfish, salamanders, frogs, and minnows.

Enemies. Raccoons are probably an important predator, especially when queen snakes are lethargic in cool weather. Otters also probably eat a few individuals in the larger streams.

In Captivity. Queen snakes generally make poor captives. Feeding is usually a problem unless there is a creek nearby with a large crayfish population. Even then many queen snakes refuse to eat. Temperament is also variable—some are docile, others are irascible.

Folklore. The long-lived but false belief that this species is poisonous is currently seen in the "Peanuts" cartoon strip where Lucy and Charlie Brown live in fear of the "deadly Queen Snake."

Location of Specimens. AMNH, CHM, CM, CWM, FMNH, GMU, INHS, LC, MCZ, MLBS, NVCC, OSU, UAMNH, UK, UMMZ, USFWS, USNM, VCU, VMNH.

Glossy Crayfish Snake
(*Regina rigida*)

Other Common Names: Swamp snake, stiff snake. *Plate 7*

Description. Adult: Brown to olive-brown above, sometimes with two faint parallel stripes. Upper lip is yellow. Scales are very shiny, making this the shiniest of the water snakes. Belly is yellow to yellowish-brown with two rows of distinct black spots or half-moons which converge in

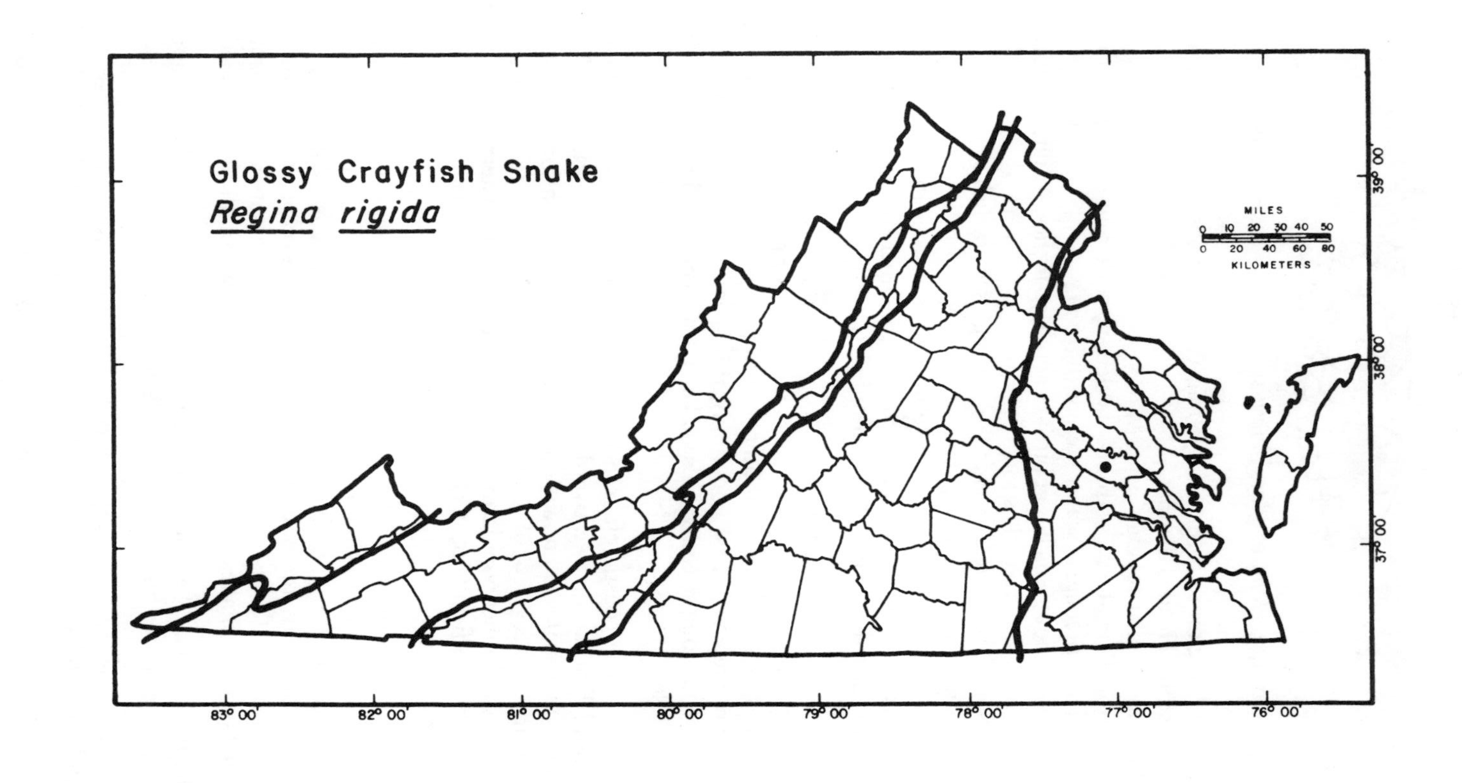
Glossy Crayfish Snake
Regina rigida
MILES
0 10 20 30 40 50
0 20 40 60 80
KILOMETERS
39° 00'
38° 00'
37° 00'
83° 00'
82° 00'
81° 00'
80° 00'
79° 00'
78° 00'
77° 00'
76° 00'

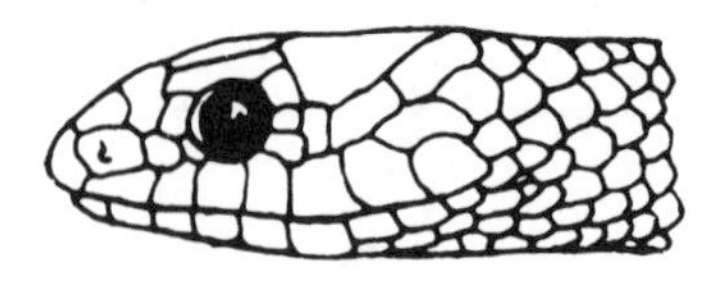

Fig. 12. Head of glossy crayfish snake, 3× natural size

the neck region to form a single mid-ventral dark stripe. Body is moderately stout. Head is distinct.
Juvenile: Similar to adult.
Scalation: Dorsal scales strongly keeled; 19 scale rows; anal plate divided. Loreal and preocular scales present (fig. 12).
Size: From about 7 inches at birth to about 30 inches. Adults are normally between 12 and 24 inches long.
Variation: *Regina rigida rigida* (Say), the glossy crayfish snake, is the only subspecies occurring in the state.
Similar Species: The queen snake has stripes rather than large spots on the belly, a slimmer body, and a small head. Other species of water snakes usually have dorsal blotches or a red belly.

Habitat. Throughout its southern range, the glossy crayfish snake is a serpent of the lowlands. It inhabits the muddy or mucky edges of swamps, marshes, ponds, lakes, and other still waters. Areas with aquatic vegetation are preferred. Less commonly, this species has been found in open water and along streams. This snake may also be found in brackish water areas (Neill, 1958).

Range. This coastal plain species ranges from eastern Virginia to north-central Florida and westward to eastern Texas and Oklahoma. In Virginia, this snake is known only from a small population in New Kent County. Huheey (1959), Hardy (1972), and Musick (1972) make reference to a disjunct population of this species in southeastern Virginia. This species has officially been classified as "Status Undetermined" in Virginia (Mitchell, 1991).

Habits. Glossy crayfish snakes are the most secretive of all Virginia water snakes. They hide in mud or muck, under debris in or near water, and under cover of aquatic plants. Most of their time is spent in the water, with individuals rarely coming into the open except at night or when driven out by floods. Because of their secretive habits and inhospitable habitat, these snakes are among the least known of all Virginia species.

Reproduction. Females bear living young. Newborn snakes are approximately 7 inches long at birth. Little else is known.

Food. The food of this species consists of small fish, salamanders, frogs, and crayfish.

Enemies. These snakes are preyed upon by predators similar to those attacking other small, semiaquatic snakes.

In Captivity. Glossy crayfish snakes should be kept in a semiaquatic terrarium that includes aquatic plants and shelter, both in and out of the water.

Location of Specimens. CM.

BROWN SNAKES
(Genus *Storeria*)

Brown snakes are small, slender, secretive, terrestrial snakes. The dorsal scales are keeled, and the anal plate is divided. These snakes are found in areas ranging from city parks and backyards to mountain wilderness areas. They feed on a variety of invertebrates including earthworms, insects, slugs, snails, millipedes, centipedes, and sowbugs. Newly transformed frogs and toads may also be taken if available. When alarmed, brown snakes may flatten their bodies and use their anal scent glands in the same manner as water snakes. Brown snakes are ovoviviparous. Up to 24 young may be born from June to mid-September. Newborn individuals are generally between 3 and 4 inches long.

Two members of this genus are found in Virginia: the northern brown snake and the northern red-belly snake. Some evidence of intergradation between the northern brown snake and the midland brown snake is seen in the extreme southeastern and southwestern sections of the state. The latter's influence can be seen in certain individuals displaying many dark lines across the back and 176 or more ventral and subcaudal scales.

Northern Brown Snake
(*Storeria dekayi*)

Other Common Names: Ground snake, Dekay's snake. *Plates 8 and 9*

Description. Adult: Yellowish- or reddish-brown to dark brown with two parallel rows of black spots down the back. Occasionally the spots may be connected by light lines across the back, especially in the southern extreme of Virginia. Additional dark spots may be found on the sides. The belly is yellowish-brown with dark dots along the outer edges. A dark, nearly vertical streak is found behind the eye.
Juvenile: Body is grayish-brown, usually darker than adults, with little evidence of spots. A yellowish collar is present on the neck.
Scalation: Dorsal scales keeled; 17 scale rows; anal plate divided. Preocular scale present; loreal scale absent (fig. 13).

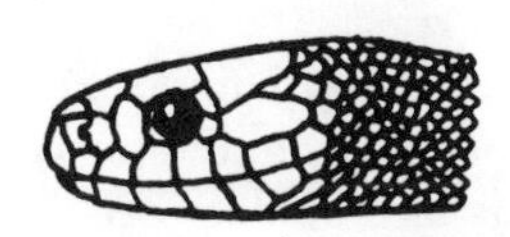

Fig. 13. Head of northern brown snake, 3× natural size

Size: From about 4 inches at birth to nearly 20 inches. Adults are usually about 12 inches in total length.
Variation: Two subspecies may be recognized in Virginia.

Plate 8

Storeria dekayi dekayi (Holbrook). Northern Brown Snake. Paired dorsal spots not usually connected by narrow light lines; 175 or fewer ventral and subcaudal scales. Statewide.

Plate 9

Storeria dekayi wrightorum Trapido. Midland Brown Snake. Paired dorsal spots usually connected by narrow light lines; 176 or more ventral and subcaudal scales. Extreme southeastern Virginia populations may show characteristics of this subspecies which is found in states to the south and west of Virginia.
Similar Species: See Tables E and F above.

Habitat. Damp areas with plenty of ground cover such as logs, boards, and rocks are prime brown snake habitats. These snakes are often found in suburban yards and even urban parks. Moist woodlands, meadows, and the edges of swamps and marshes are typical habitats in wilder areas.

Range. Brown snakes are found from Maine, Wisconsin, Minnesota, and southern Canada south to the Gulf Coast and into Mexico. Their range extends westward to South Dakota, Kansas, Oklahoma, and Texas. These snakes may be found throughout Virginia.

Habits. Like the other small species of snakes in Virginia, the northern brown snake is quite secretive during the day. At night, however, these reptiles roam their terrestrial environment in search of prey. The brown snake's eyes are large, especially in comparison with the ringneck, worm, and earth snakes with which they are often confused. This is an indication of the fact that the brown snake is not so subterranean in its habits as the other three snakes.

Reproduction. Females bear living young. Newborn brown snakes are between 3½ to 4½ inches long and are born in late July and August after a gestation period averaging 109.5 days (range 105–113 days) (Clausen, 1936; Fitch, 1970). Litters may consist of 20 or more individuals, but usually less. Newborn snakes are very dark or nearly black in appearance with a grayish-yellow band around the neck.

Food. Earthworms and slugs are the favorite foods. Snails, soft-bodied insects, and insect larvae are also eaten. Although these various invertebrates make up the bulk of the diet, small tree frogs may also be preyed upon.

Enemies. Brown snakes often end up in the stomachs of nocturnal prowlers like raccoons, opossums, and skunks. Kingsnakes and racers feed upon them, and toads and spiders eat the young. In many areas, house cats toy with these snakes, but often do not eat them.

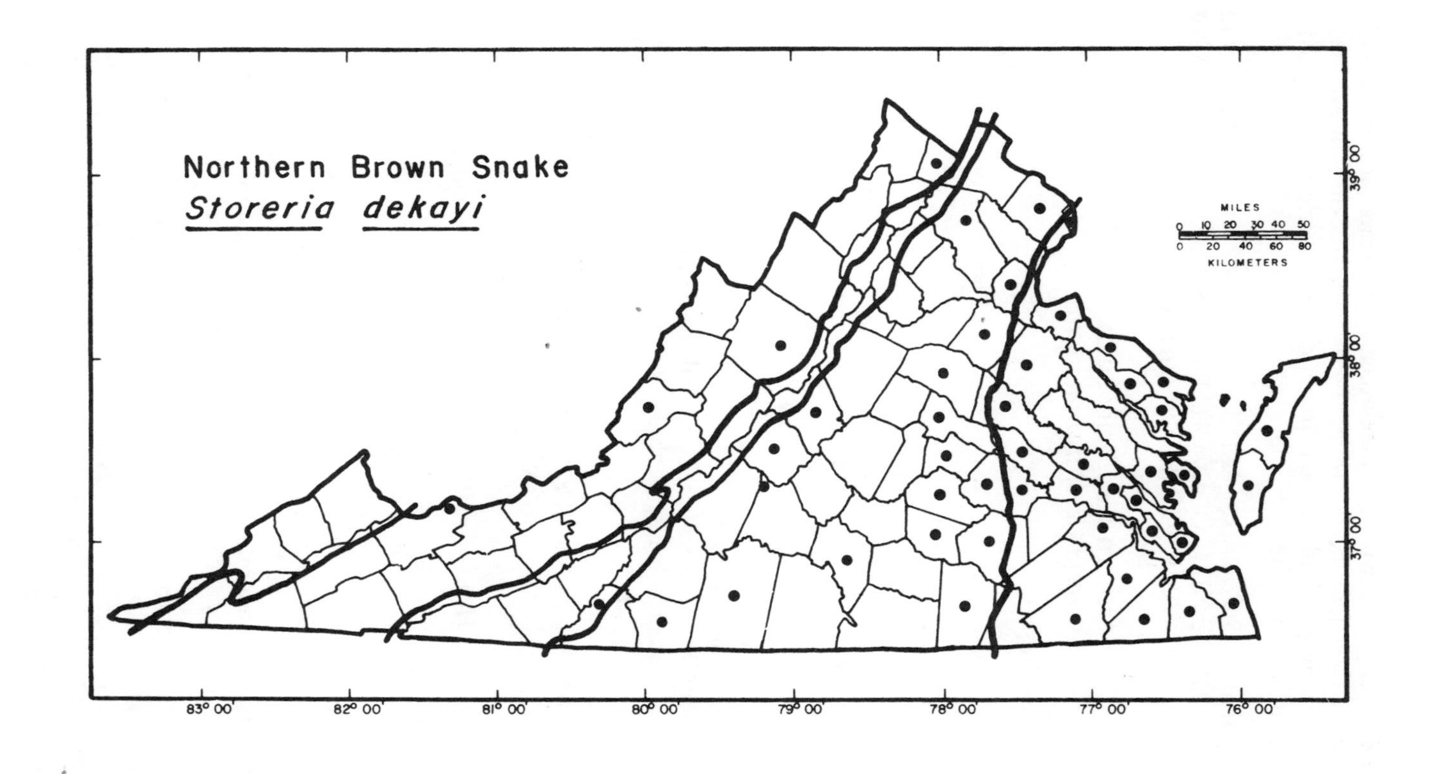
Northern Brown Snake
Storeria dekayi
MILES
0 10 20 30 40 50
0 20 40 60 80
KILOMETERS
39° 00'
38° 00'
37° 00'
83° 00'
82° 00'
81° 00'
80° 00'
79° 00'
78° 00'
77° 00'
76° 00'

In Captivity. Brown snakes are excellent residents for a woodland terrarium. Their food needs are met by periodic stockings of earthworms. The terrarium should not be kept too damp. Water can be supplied in a small upturned jar lid.

Location of Specimens. AMNH, CFR, CM, CWM, DS, DU, ESU, GMU, GMW, NLU, ODU, OU, PNSC, SDNHM, UMMZ, MSWB, USNM, VCU, VIMS, VMNH, VPI and SU.

Northern Red-Belly Snake (*Storeria occipitomaculata*)

Plate 10

Other Common Names: Worm snake, red-bellied ground snake, spot-necked snake.

Description. Adult: Usually brown above, but may vary from gray to nearly black. Four indistinct dark stripes and/or a broad light stripe may occur on the back. Belly is usually bright red with no markings. Occasional individuals may be orange, pink, yellow, or rarely bluish-black below. Three pale spots, which may connect to form a collar, are usually present on the neck.
Juvenile: Similar to adult.
Scalation: Dorsal scales keeled; 15 scale rows; anal plate divided. Preocular scale present; loreal scale absent (fig. 14).

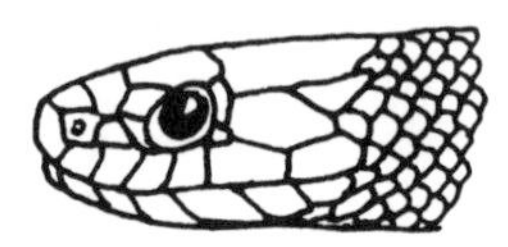

Fig. 14. Head of northern red-belly snake, 3× natural size

Size: From about 3 inches at birth to 16 inches. Adults are usually between 8 and 10 inches in total length.
Variation: *Storeria occipitomaculata occipitomaculata* (Storer), the northern red-belly snake, is the only subspecies occurring in the state.
Similar Species: See Tables E and F above. Other small snakes with neck markings have smooth scales, except for the northern brown snake (juvenile) which has 17 scale rows and a belly which is never reddish.

Habitat. Northern red-belly snakes are usually found in wooded areas, especially those with rocks, logs, or other cover. Habitats may be moist or dry, ranging from the edges of swamps and bogs to well-drained, rocky hilltops. The debris around old barns and abandoned houses often support colonies of these little snakes.

Range. Red-belly snakes are found from Maine and southern Canada south to central Florida and west to eastern Texas, Oklahoma, Kansas,

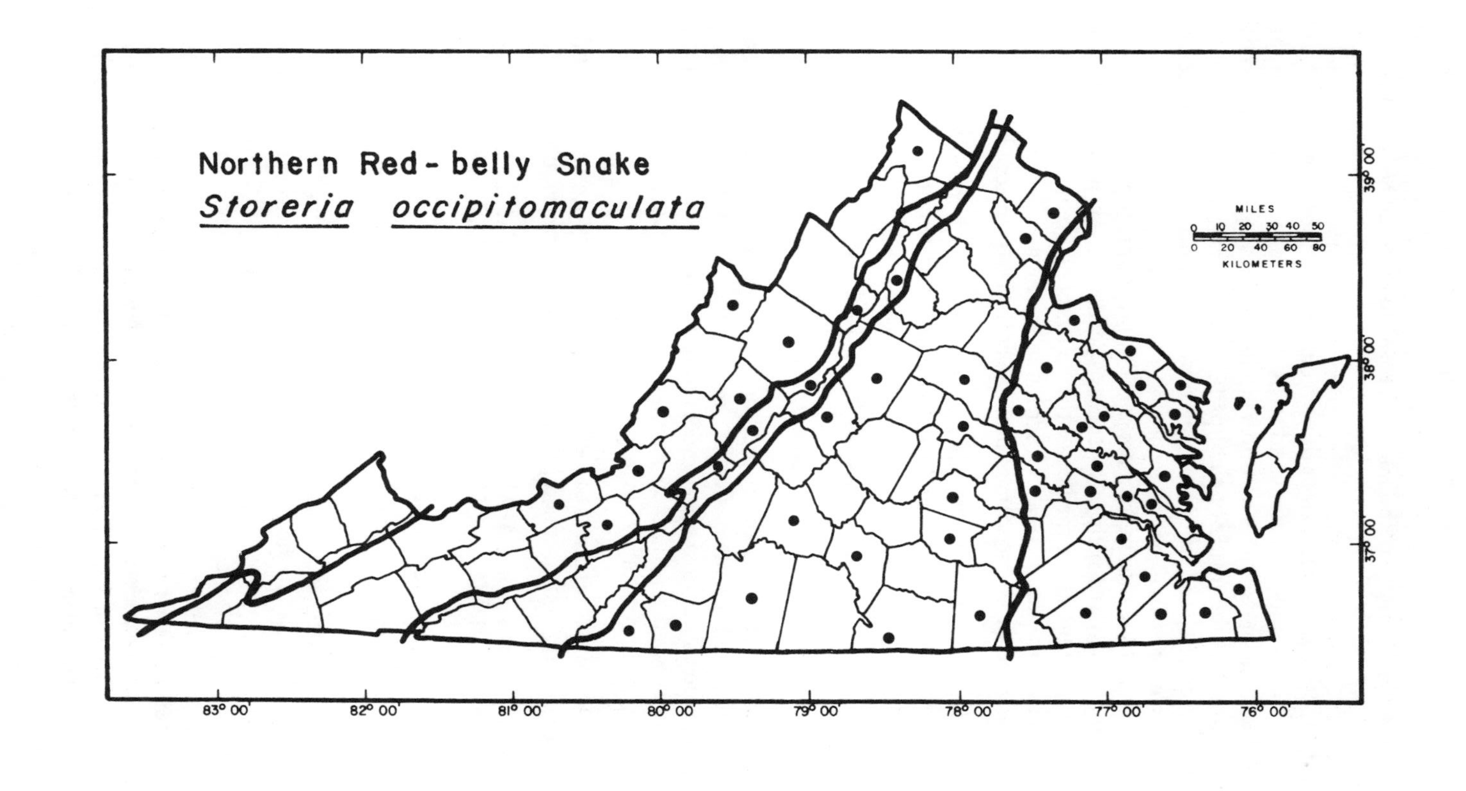
Northern Red-belly Snake
Storeria occipitomaculata
MILES
0 10 20 30 40 50
0 20 40 60 80
KILOMETERS
39° 00'
38° 00'
37° 00'
83° 00'
82° 00'
81° 00'
80° 00'
79° 00'
78° 00'
77° 00'
76° 00'

South Dakota, and North Dakota. These snakes may be found throughout Virginia.

Habits. Some individuals of this species demonstrate some unusual defensive habits under certain conditions. Lip curling is a reaction to threat in which the snake raises its upper lip like a snarling dog, thus exposing the gums and the inside of the lips (Gosner, 1942). The snake does not snarl, of course, and what is achieved by this is uncertain. Death feigning or "playing possum" has also been observed. In one case, a female snake played dead only during the time she was pregnant. A physical characteristic, the bright red belly, also causes speculation on its possible use in defense, perhaps along the line of the "corkscrew" reaction used by ringneck snakes in certain areas. Other habits of the northern red-belly snake are similar to those mentioned previously for the northern brown snake.

Reproduction. Females bear living young. Up to 21 young are born during the summer, usually in July and August. Newborn snakes vary in length from approximately 2½ inches to 4 inches.

Longevity. Maximum known age: 2 years, 2 months (Bowler, 1977).

Food. Insects, earthworms, slugs, and occasionally tiny frogs are eaten.

Enemies. The animals that prey on this species are much the same as those that feed on various other small snakes. One unusual record occurred when a marbled salamander (*Ambystoma opacum*) killed and partially swallowed an 8½ inch northern red-belly snake. This occurred in captivity and probably rarely happens in the wild.

In Captivity. Northern red-belly snakes generally do well in captivity under conditions mentioned previously for the northern brown snake.

Location of Specimens. AMNH, CFR, CM, CWM, DU, GMW, HSH, MCZ, NVCC, UMMZ, USNM, VCU, VMNH, VPI and SU.

GARTER SNAKES
(Genus *Thamnophis*)

Garter snakes and their relatives are among the most widely distributed snakes in the United States with varieties occurring from coast to coast and from border to border. They range north in Canada and south through Mexico. Two members of this group occur in Virginia: the eastern garter snake and the eastern ribbon snake. A western subspecies of our eastern garter snake is the only snake found in Alaska.

Garter snakes are easily distinguished from other Virginia snakes because of their light longitudinal stripes. In some individuals only a single mid-dorsal stripe is present, but usually an additional stripe is present on each side. In all cases the stripe is lighter than the background coloration. The common name of this group is derived from the resemblance of this pattern to an old-fashioned striped garter. In addition to their distinguishing pattern, these snakes possess keeled dorsal scales and a single, undivided anal plate.

Garter snakes inhabit a wide variety of areas and feed primarily on frogs, toads, salamanders, and earthworms. Females are viviparous and give birth to as many as 80 or more young from mid to late summer. Newborn snakes are normally between 6 and 9 inches long.

Eastern Garter Snake (*Thamnophis sirtalis*)

Plates 11–13

Other Common Names: Garden snake, grass snake, striped snake.

Description. Adult: Coloration and pattern variable, but usually with a combination of stripes and spots. Usual pattern consists of three stripes which are generally yellow but which may have a bluish or greenish tinge. The lowermost stripes involve the 2d and 3d scale rows along both sides of the body. Between the stripes may be two rows of dark spots on a background of olive, brown, or green. Sometimes the spots are absent and the area between the stripes is dark brown or black with light flecks. Belly is greenish-yellow with a row of black spots on each side of the ventrals. Body is moderately slender. Tail comprises less than 1/4 of total length.
Juvenile: Similar to adult.
Scalation: Dorsal scales keeled; 19 scale rows; anal plate undivided. Loreal and preocular scales present (fig. 15).

Fig. 15. Head of eastern garter snake, natural size

Size: From about 6 inches at birth to 48 inches. Most adults are between 18 and 30 inches long.
Variation: *Thamnophis sirtalis sirtalis* (Linnaeus), the eastern garter snake, is the only subspecies occurring in the state.
Similar Species: The eastern ribbon snake is more slender than the eastern garter snake, has a tail which is more than 1/4 of the total length, and its lateral stripes occupy the 3d and 4th scale rows (2d and 3d scale rows in the eastern garter snake). Yellow stripes on the queen snake occur only on the sides. The three stripes on the rainbow snake are dark red. Stripes on brown snakes, red-belly snakes, and canebrake rattlesnakes are *darker* than the background coloration.

Habitat. A wide variety of moist environments contain populations of this species. Weed patches, orchards, meadows, gardens, and open woods are examples. Marshes (freshwater and brackish), swamps, streamsides, pond shores, and similar wet areas also support numbers of garter snakes. This is a very adaptable species that is just as likely to be found in some suburban garden (hence one of its common names—"garden snake") as in a deep wilderness.

Range. Garter snakes are found throughout most of the United States. They occur statewide in Virginia.

Habits. Eastern garter snakes are alert, active snakes that may be on the prowl either day or night. Basically terrestrial, they seldom climb but may take to water. When attempting to escape from an enemy, these snakes rely on speed and the optical illusion created by the dorsal stripe. An attacker's eyes focus on the stripe which appears stationary even though the snake is quickly gliding away. When cornered, most eastern garter snakes will flatten their body and strike, but after being caught their reactions vary. Some are docile; others bite vigorously.

Eastern garter snakes are probably the most cold-tolerant of all Virginia snakes and have even been observed crawling over patches of snow. They are usually the last species to enter hibernation in the fall and the first to emerge in the spring. Some individuals may emerge as early as February.

Reproduction. Females can reach sexual maturity by the second spring after birth (Carpenter, 1952) at which time they normally are 20 inches or more in total length. Between 3 and 85 (average 16–18) living young are born from mid to late summer (Hoffman, 1970). Wood and Wilkinson (1952) recorded a litter of 59 garter snakes born on July 30 in Warwick County (Newport News). Newborn individuals range between 5 and 8 inches in total length. Generally, the larger females produce the greater number of young. Dunlap and Lang (1990) presented data showing that clutches of large females were male-biased whereas those of small females were female-biased.

Longevity. Maximum known age: 10 years (Bowler, 1977).

Food. Frogs, toads, salamanders, and earthworms are favorite foods, although eastern garter snakes have also been known to eat mice, young birds, snakes, fish, insects, and leeches. They are extremely voracious and will sometimes attempt to swallow a handler's finger! Uhler, Cottam, and Clarke (1939) examined 24 garter snake stomachs taken from the George Washington National Forest in Virginia. Major food items, by volume, included earthworms 37%, salamanders 31%, toads 25%, and snakes 4%.

Enemies. Predators take a heavy toll of garter snakes, but the large numbers of young born each year offset these losses. Raccoons, opossums, skunks, weasels, hawks, owls, kingsnakes, black racers, and even bull-

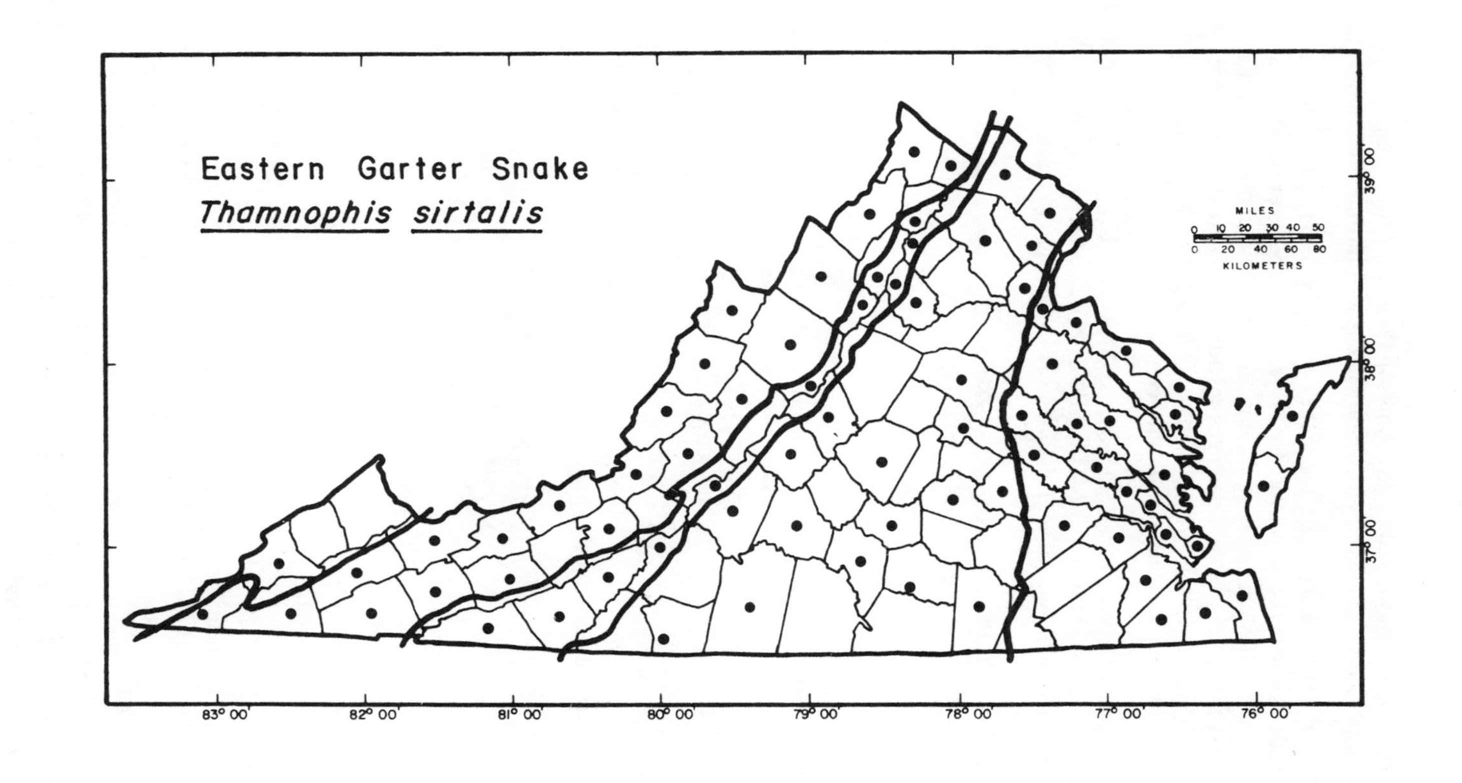
Eastern Garter Snake
Thamnophis sirtalis
MILES
0 10 20 30 40 50
0 20 40 60 80
KILOMETERS
39° 00'
38° 00'
37° 00'
83° 00'
82° 00'
81° 00'
80° 00'
79° 00'
78° 00'
77° 00'
76° 00'

frogs feed on these snakes. Humans kill their share too, but many persons recognize the light stripes as a characteristic of harmlessness.

In Captivity. Eastern garter snakes are ready feeders, and the variety of foods eaten make this task doubly easy for the keeper. Various store-bought items such as bologna, hot dogs, and fish can be tried if natural foods are not available. Garter snakes seem to have less ability to fast than most other species, however. Many individuals become docile in captivity; others remain nervous and ill-tempered.

Location of Specimens. AMNH, ANSP, CM, CU, CWM, DU, FMNH, GMU, GMW, HSH, LC, MCZ, MLBS, NVCC, ODU, SDSNH, TAMU, UAMNH, USFWS, USNM, VCU, VIMS, VMNH, VPI and SU, WCC.

Eastern Ribbon Snake
(*Thamnophis sauritus*)

Plate 14

Other Common Names: Slender garter snake, striped water snake, striped racer.

Description. Adult: Dorsal pattern of three bright yellow stripes on a dark background. The lowermost stripes involve the 3d and 4th scale rows along both sides of the body. Plain yellow or greenish-yellow belly. Body very slender with the long tail comprising more than 1/4 of the total length.

Juvenile: Similar to adult.

Scalation: Dorsal scales keeled; 19 scale rows; anal plate undivided. Loreal and preocular scales present (fig. 16).

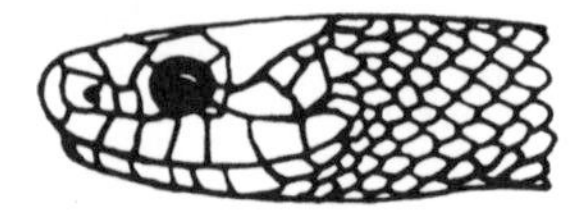

Fig. 16. Head of eastern ribbon snake, 2× natural size

Size: From about 7 inches at birth to about 36 inches. Most adults are about 2 feet long.

Variation: *Thamnophis sauritus sauritus* (Linnaeus), the eastern ribbon snake, is the only subspecies occurring in the state.

Similar Species: The eastern garter snake is more heavy-bodied, has a tail which is less than 1/4 of the total length and has lateral stripes occupying the 2d and 3d scale rows. No other striped snakes have a yellowish mid-dorsal stripe.

Habitat. The shallow waters and shorelines of creeks, ponds, swamps, marshes (freshwater and brackish), and wet meadows are the eastern ribbon snake's habitat. The preferred areas are usually well-vegetated with

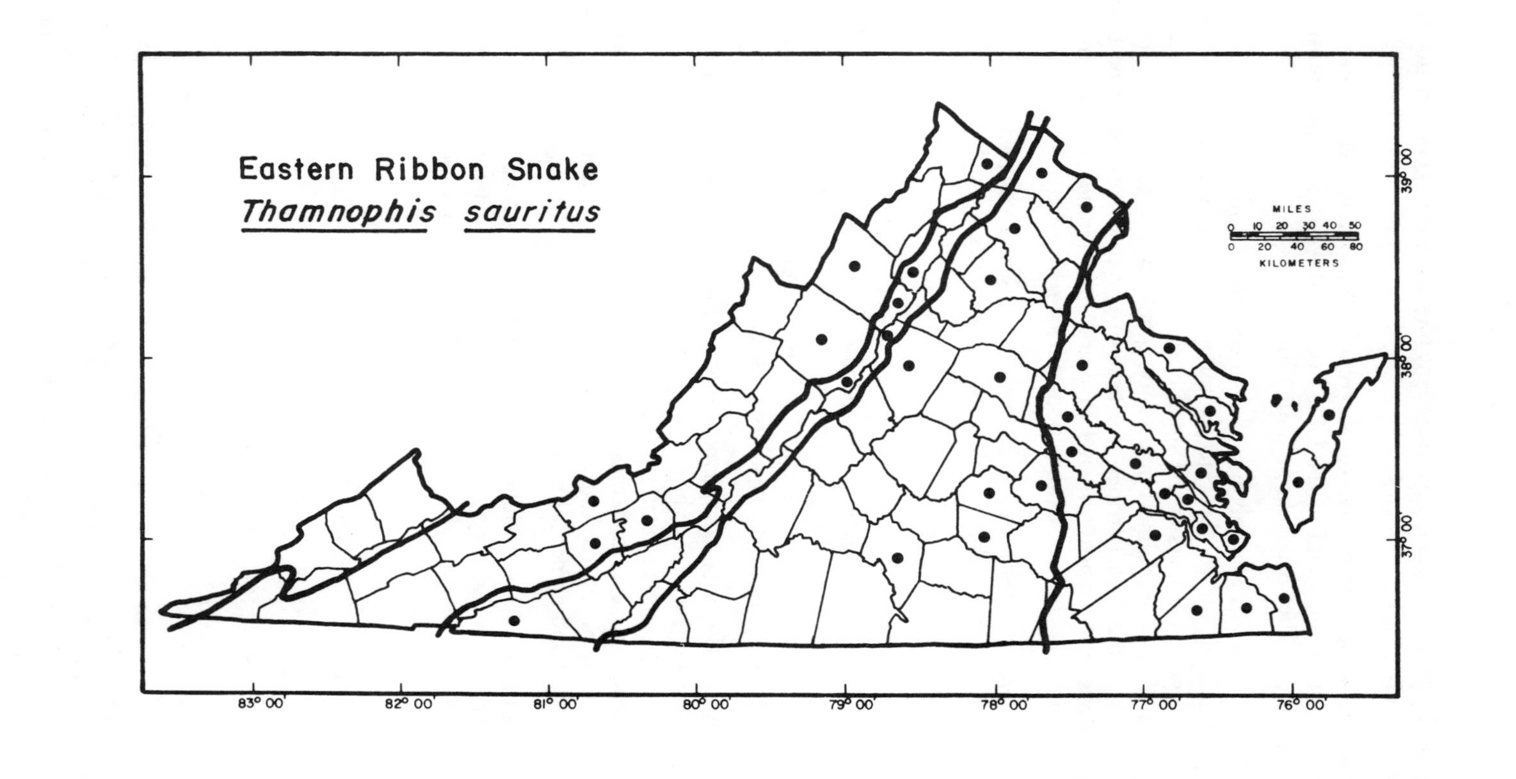

Eastern Ribbon Snake
Thamnophis sauritus
MILES
0 10 20 30 40 50
0 20 40 60 80
KILOMETERS
39° 00'
38° 00'
37° 00'
83° 00'
82° 00'
81° 00'
80° 00'
79° 00'
78° 00'
77° 00'
76° 00'

cattails, grasses, shrubs, or other plant life that offers good traction for quick escapes.

Range. Eastern ribbon snakes are found from southern Maine, Pennsylvania, and Ohio south to the Gulf Coast and westward into Louisiana and Illinois. They occur throughout Virginia.

Habits. The eastern ribbon snake is a semiaquatic species with racerlike speed on land and good surface swimming ability in water. These snakes are also good climbers in the small bushes along the water's edge. They are active, nervous animals that rely on quickness and the optical illusion created by the dorsal stripes (see eastern garter snake) to escape from predators. Even when cornered they coil only momentarily before attempting to dart off in the direction of least resistance.

Reproduction. Females give birth to as many as 20 living young in mid to late summer. Newborn snakes are between 7 and 9 inches long.

Longevity. Maximum known age: 3 years, 11 months, 29 days (Bowler, 1977).

Food. Frogs, salamanders, toads, small fish, and leeches are eaten. Two snakes examined by Uhler, Cottam, and Clarke (1939) from the George Washington National Forest had fed upon a cricket frog and harvester ants. All prey is normally swallowed alive.

Enemies. The usual complement of shoreline and shallow-water predators feed on eastern ribbon snakes. Raccoons, otters, mink, herons, kingsnakes, snapping turtles, bullfrogs, bass, and pickerel will take their share when they can catch these speedy snakes. As with the eastern garter snake, many people recognize the striped pattern as indicative of a harmless species and do not kill eastern ribbon snakes.

In Captivity. Eastern ribbon snakes remain high-strung and nervous in captivity. However, they generally feed well if provided with a suitable hiding place. One female ate frogs the day before and the day after she gave birth to 10 young.

Location of Specimens. CRF, CM, GMU, GMW, LC, MCZ, MLBS, NVCC, ODU, SDSNH, UMMZ, USFWS, USNM, VCU, VIMS, VMNH.

EARTH SNAKES
(Genus *Virginia*)

Earth snakes are very small snakes that rarely grow longer than 12 inches. Their name is well deserved for both their habitat and coloration are "earthy." These snakes have a moderately stout body and a relatively short tail. The anal plate is divided, and the dorsal scales may be either smooth or keeled.

Two species of earth snakes are found in Virginia—the smooth earth snake and rough earth snake. Both of these snakes are secretive and are often found under piles of leaves, in rotting logs and stumps, and by persons digging in gardens or plowing fields. They feed primarily on earthworms, snails, slugs, and insects. Earth snakes breed during late spring or early summer with females giving birth to a maximum of approximately 12 young in late summer. Newborn snakes are 4 to 5 inches long.

Smooth Earth Snake
(*Virginia valeriae*)

Plate 15

Other Common Names: Brown snake, worm snake, ground snake.
Description. Adult: Usually medium brown, but may be gray or reddish-brown. Tiny scattered black dots and faint light lines through the scales may be present dorsally. Belly is cream-colored and unmarked. Body is moderately stout; head is wider than neck. Snout rounded.
Juvenile: Similar to adult.
Scalation: Dorsal scales smooth or weakly keeled; 15–17 dorsal scale rows; anal plate divided; 6 upper labial scales. Loreal scale present; preocular scale absent (fig. 17).

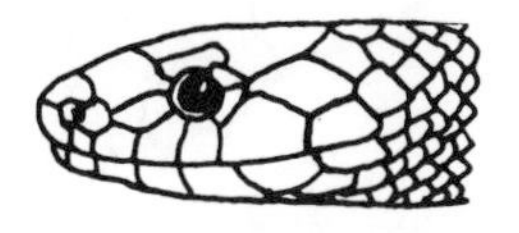

Fig. 17. Head of eastern smooth earth, 3× natural size

Size: From about 4 inches at birth to slightly over 12 inches. Normal adult size is 8 to 10 inches.
Variation: Two subspecies are recognized in Virginia.

Plate 15

Virginia valeriae valeriae Baird and Girard. Eastern Smooth Earth Snake. Dorsal scales smooth except for faint keels on parts of tail. 15 scale rows at midbody. Occurs statewide.
Virginia valeriae pulchra. Mountain Earth Snake. Dorsal scales weakly keeled. 15 scale rows on front portion of body; 17 scale rows at midbody and posteriorly. Recorded only from Highland County (Mitchell, 1974). Cervone (1983) discussed the natural history of this subspecies.
Similar Species: Rough earth snake has keeled dorsal scales, 17 scale rows, and 5 upper labial scales. Northern brown snake and worm snake may also be confusing. See Table F above.

Habitat. Smooth earth snakes seem to be most abundant in rich deciduous woodlands, although they may be found in fields, pastures, or gardens where there is adequate cover and loose soil in which to burrow. Rotting

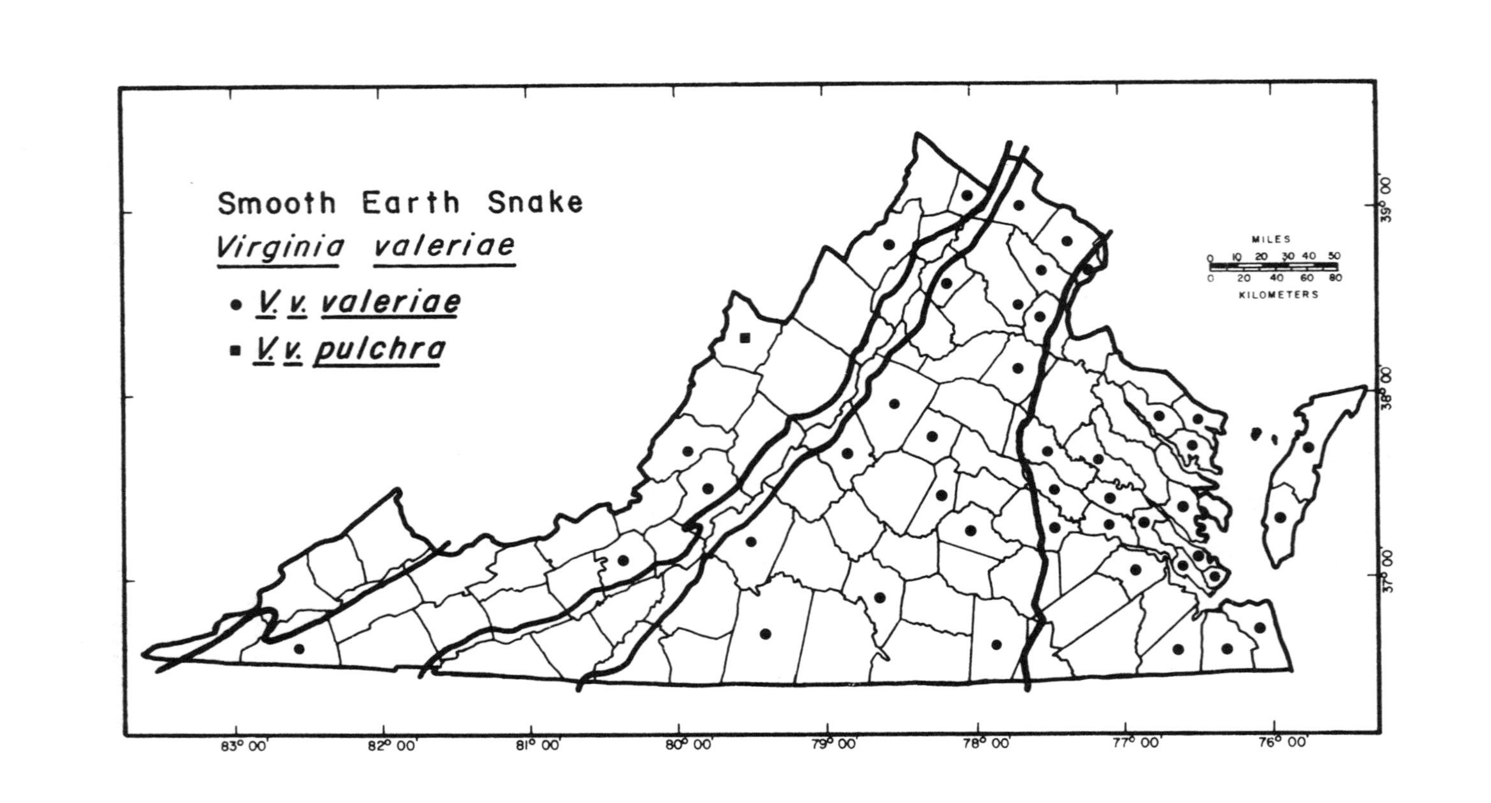
Smooth Earth Snake
Virginia valeriae
• V. v. valeriae
▪ V. v. pulchra
MILES
0 10 20 30 40 50
0 20 40 60 80
KILOMETERS
39° 00'
38° 00'
37° 00'
83° 00'
82° 00'
81° 00'
80° 00'
79° 00'
78° 00'
77° 00'
76° 00'

logs and stumps are favorite shelters, as are the boards and trash found near abandoned homesites and barns.

Range. Smooth earth snakes range from New Jersey, Pennsylvania, Ohio, Indiana, Illinois, and Iowa south to northern Florida and the Gulf Coast. The range extends westward into Texas, Oklahoma, and Kansas. This species occurs statewide.

Habits. This species usually stays underground, or at least under cover. Heavy rains during the summer will bring them out, and they are often run over by cars at such times. The cylindrical body and small eyes are typical of snakes with subterranean habits. Defensively, they rely on their secretive habits and camouflage. When picked up, an earth snake may wrap around a finger but will seldom, if ever, bite.

Reproduction. Up to 12 young, each approximately 4 inches long, are born in late summer. Blem and Blem (1985) recorded a mean litter size of 6.6 for five litters in Virginia.

Longevity. Cervone (1983) reported life spans of 6 to 7 years or more.

Food. Earthworms, small slugs, snails, soft-bodied insects, and insect larvae are swallowed alive.

Enemies. The secretive habits of smooth earth snakes make them less susceptible to many predators. Mole snakes, scarlet snakes, milk snakes, and other reptile-eating serpents undoubtedly prey on earth snakes. Raccoons and opossums find them when rooting around rotten logs and stumps. Young earth snakes are so small that they may be attacked by large spiders, certain beetles, and toads.

In Captivity. Earth snakes do well in a woodland terrarium but are not often seen on the surface. Thus, they are not particularly interesting captives. Feed them simply by maintaining a supply of earthworms and grubs in the terrarium. Be careful, however, about keeping certain other snakes in the same terrarium. Young 8-inch mole snakes have attempted to swallow 10-inch earth snakes!

Location of Specimens. AMNH, ANSP, ASU, CFR, CM, CWM, GMW, JHU, MCZ, NLU, NVCC, UMMZ, USNM, VCU, VIMS, VPI and SU.

Rough Earth Snake
(*Virginia striatula*)

Plate 16

Other Common Names: Brown snake, worm snake, ground snake.

Description. Adult: Grayish- to reddish-brown above; cream, pinkish-, or greenish-white below. Body moderately stout. Snout pointed.
Juvenile: Similar to adult.

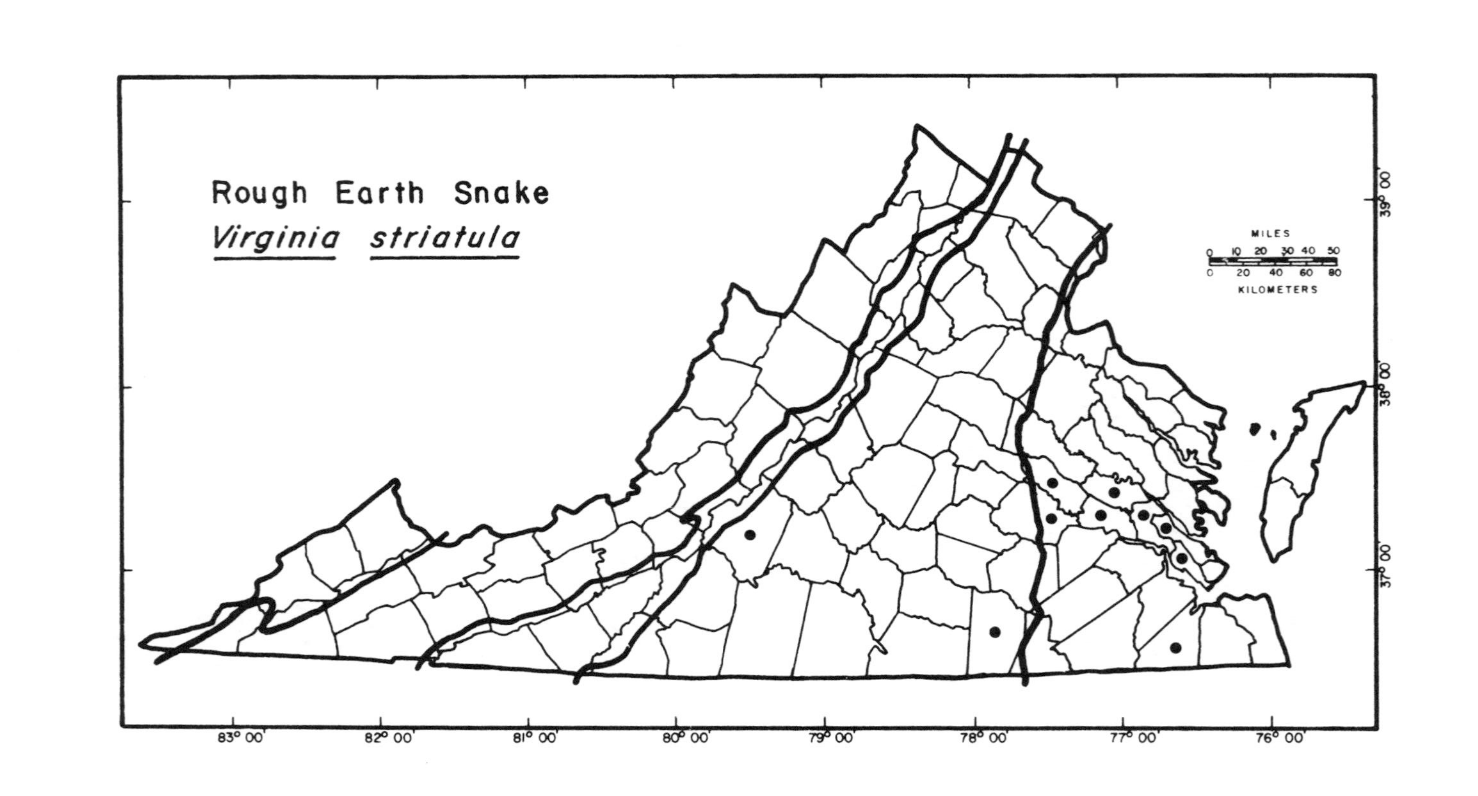
Rough Earth Snake
Virginia striatula
MILES
0 10 20 30 40 50
0 20 40 60 80
KILOMETERS
39° 00'
38° 00'
37° 00'
83° 00'
82° 00'
81° 00'
80° 00'
79° 00'
78° 00'
77° 00'
76° 00'

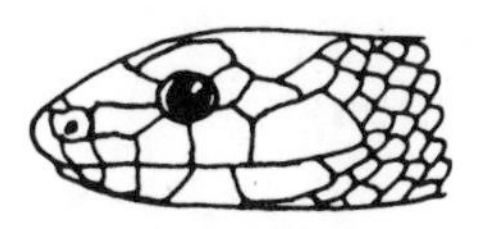

Fig. 18. Head of rough earth snake, 3× natural size

Scalation: Dorsal scales strongly keeled; 17 scale rows; anal plate divided; 5 upper labial scales. Loreal scale present; preocular scale absent (fig. 18).
Size: From about 4 inches at birth to slightly over 12 inches. Usual adult size is 8 to 10 inches.
Variation: No subspecific variation is recognized.
Similar Species: Eastern smooth earth snake has mostly smooth dorsal scales, 15 scale rows, and 6 upper labial scales. Mountain earth snake has weakly keeled dorsal scales, 15–17 scale rows, and 6 upper labial scales. May also be confused with the northern brown snake and the worm snake. See Table F above.

Habitat. Rough earth snakes live among decaying logs and rotting stumps in woodlands and along field edges. They also apparently spend considerable time underground in loose soil. Evidence indicates that this species prefers wetter areas than does the smooth earth snake. Occasionally, rough earth snakes are found hidden under cover in suburban areas.

Range. The rough earth snake ranges from southeastern Virginia to northern Florida and west into Texas, Oklahoma, Kansas, and Missouri. In Virginia, this snake has been taken almost exclusively in the southeastern one-third of the state. One record exists, however, from Bedford County.

Habits. This species is seldom seen because of its secretive habits. Heavy rains will bring them into the open. At other times they hide under various kinds of shelter and underground. Rough earth snakes emerge from hibernation early. They have been found in hibernating masses with copperheads, ribbon snakes, lizards, frogs, and toads. A group of at least 30 rough earth snakes were accidentally excavated from a hibernaculum at Richmond on March 30, 1982 (Blem and Blem, 1985).

Reproduction. Females in Virginia probably give birth in late July or early August. Mean litter size is 6.0 (range 4–10) for 24 litters (Blem and Blem, 1985). Newborn snakes are approximately 4 inches long.

Food. Earthworms, snails, ant eggs, and soft-bodied insects are eaten alive.

Enemies. Similar to smooth earth snake.

In Captivity. Similar to smooth earth snake.

Location of Specimens. CM, PNSC, UK, UMMZ, USNM, VCU, VPI and SU.

HOGNOSE SNAKES (Genus *Heterodon*)

One of the most interesting snakes in Virginia is the hognose snake. These snakes are spectacular bluffers whose tricks include flattening the neck, swelling with air, hissing, and feigning death. Despite their frightening behavior, however, they are completely harmless to man.

Hognose snakes possess keeled dorsal scales, a divided anal plate, and a very distinctive upturned snout. A pair of enlarged teeth are present in the posterior portion of the upper jaw in hognose snakes. Faint traces of an anterior groove are present on these teeth. These grooves are similar to the grooves found in certain poisonous snakes. The enlarged teeth are probably used primarily for manipulating and deflating toads before swallowing them. The salivary secretions of hognose snakes have been found to be toxic to some amphibians and mildly venomous to humans (Bragg, 1960; McAlister, 1963; Grogan, 1974). The saliva is hemotoxic and causes local edema and hemorrhaging of blood vessels.

Eastern Hognose Snake (*Heterodon platirhinos*)

Plates 17–21

Other Common Names: Spread-head moccasin, spreading adder, blowing viper, puff viper, possum snake, hissing adder, puff head.

Description. Adult: Extremely variable. Usually a row of dark dorsal blotches alternating with a row of dark lateral blotches. Colors include black, brown, gray, orange, red, and yellow. Some or all of these colors may be present on one individual. Solid black adults are not rare, comprising 10 to 20 percent or more of some populations. Solid gray, brown, and red specimens have been reported. Belly is usually greenish-yellow or gray and is mottled with darker gray or brown. Upturned, pointed snout is the best distinguishing characteristic.
Juvenile: Similar to adult.
Scalation: Dorsal scales keeled; 23–25 scale rows; anal plate divided. Loreal and preocular scales present (fig. 19).

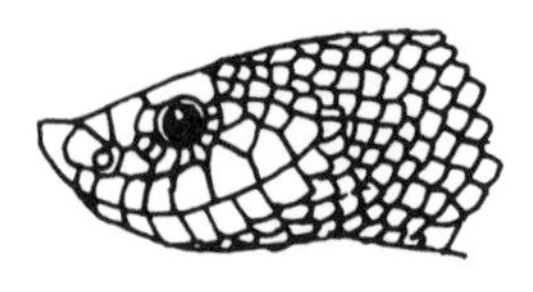

Fig. 19. Head of eastern hognose snake, natural size

Size: From about 7 inches at hatching to over 42 inches. Most adults are between 18 and 30 inches in total length.
Variation: No subspecific variation is recognized.

Similar Species: Copperhead and other blotched snakes (see Table D above). Black hognose snakes may be confused with other black-colored species (see Table C above).

Habitat. The presence of sandy or sandy-loam soil is apparently the key ingredient of hognose snake habitat. In these areas they may be found in open hardwoods, pine stands, or fields. They prefer well-drained land to wet areas, but may be found in the vicinity of ponds. They do not seem to be particularly attracted to farm buildings. Hardy and Olmon (1971) noted two individuals in brackish water. One of these was swimming in over nine fathoms of saline water (21 ppt) one-half mile offshore in the York River, Virginia.

Range. Eastern hognose snakes range from Massachusetts, New York, Ontario, Michigan, Wisconsin, and Minnesota south to the Gulf Coast and southern Florida. The range extends westward to central Texas, western Oklahoma, Kansas, Nebraska, and South Dakota. They are found throughout Virginia.

Habits. The defensive habits of the eastern hognose snake are fascinating. When approached it will flatten its head and neck, inflate with air, and exhale with loud hisses. The snake will make short lunges but does not bite, for the actor is all bluff. If poked or kicked the snake will writhe in apparent agony with mouth open and tongue dragging in the dust. Feigned death soon overtakes the poor creature and it lies limply on its back. The snake overplays its role, however; when placed belly down, it quickly twists onto its back. If the attacker leaves, the hognose snake shortly lifts its head to survey the situation and moves off to safety.

This species is diurnal and follows the scent trails of toads to their daylight hiding places. The snout is used in rooting out the toads and in other burrowing. Hognose snakes are terrestrial and definitely not arboreal. They can swim well when necessary. Platt (1969) studied the natural history of this species.

Reproduction. The female hognose snake lays her ellipsoid, off-white, thin-shelled, nonadherent eggs during June or July. The eggs, which may number from 4 to 69 (usually 10 to 30), are between 3/4 and 1 1/2 inches long and between 1/2 and 1 inch in width. Clutches of eggs have been found beneath rocks and at depths of 4 to 6 inches in sandy fields and gravel deposits (Edgren, 1955). Dunn (1915*a;* 1915*b*) recorded a nest containing 34 eggs that was plowed up in a field in Nelson County. The 7- to 9-inch young hatch in August and September.

Longevity. Maximum known age: Linzey (unpubl.) received an adult hognose snake that was approximately 40 inches long on September 4, 1970. This snake lived until July 9, 1977—a period of 6 years, 10 months, and 5 days. Clifford (unpubl.) maintained a specimen in captivity for 5 years, 9 months. This snake was also an adult when captured.

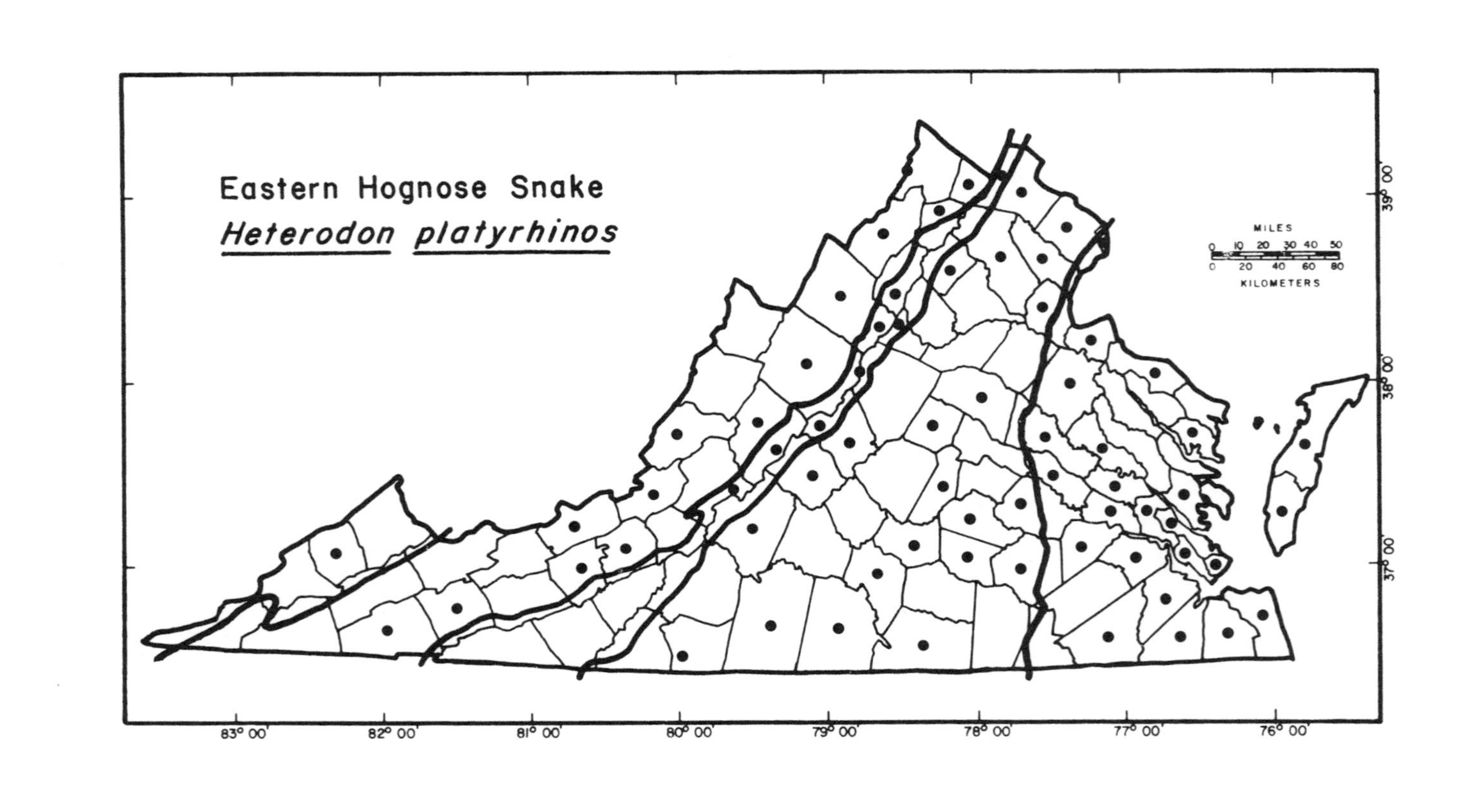
Eastern Hognose Snake
Heterodon platyrhinos
MILES
0 10 20 30 40 50
0 20 40 60 80
KILOMETERS
39° 00'
38° 00'
37° 00'
83° 00'
82° 00'
81° 00'
80° 00'
79° 00'
78° 00'
77° 00'
76° 00'

Food. Toads are the principal food, although frogs, salamanders, and insects are also eaten. Ernst and Laemmerzahl (1989) recorded a snake that had consumed a spotted salamander (*Ambystoma maculatum*) in northern Virginia. Uhler, Cottam, and Clarke (1939) examined the stomach contents of 10 hognose snakes taken in the George Washington National Forest. Major food items, by volume, were toads 40%, frogs 30%, salamanders 11%, and small mammals (mouse and chipmunk) 19%. Small birds have been reported (Edgren, 1955; Linzey, unpubl.). Young snakes feed mainly on small frogs and toads, worms, and insects. The prey is not prekilled but is eaten alive.

Enemies. The bluffing behavior of hognose snakes is probably effective against many predators, including some confrontations with humans. Inexperienced raccoons, opossums, and foxes may be stalled by it. Others, such as kingsnakes and hawks, are not fooled.

In Captivity. Eastern hognose snakes usually thrive in captivity if toads are available for food. They need a dry cage with a small bowl of water and a place to hide. One specimen, caught as a mature adult by Clifford, fed well and was still healthy after several years in captivity, although it had apparently become completely blind, probably as a result of old age.

Location of Specimens. AMNH, ANSP, CM, CU, CWM, DU, ESU, GMU, GMW, JHU, LC, MCZ, MLBS, NVCC, OU, PNSC, SDSNH, UI, UK, UMMZ, USFWS, USNM, VCU, VMNH, VPI and SU.

RINGNECK SNAKES
(Genus *Diadophis*)

Ringneck snakes are small, alert, flat-headed snakes with a distinct ring on the neck. Typically they are animals of moist woodlands that contain plenty of ground cover under which these secretive nocturnal snakes hide. They have smooth dorsal scales and a divided anal plate.

Ringneck Snake
(*Diadophis punctatus*)

Plate 22

Other Common Names: Ring snake, baby king snake, red-belly snake, yellow-belly ring snake.

Description. Adult: Gray to bluish-black above with a flat, black head. Neck ring and ventral color varies from yellowish to reddish. Neck ring may be incomplete. Belly ranges from being unmarked to having a row of dark, half-moon-shaped spots running along its length.
Juvenile: Similar to adult but with velvety black dorsum and even blacker head.

Scalation: Dorsal scales smooth; 15 scale rows; anal plate divided. Loreal and preocular scales present (fig. 20).

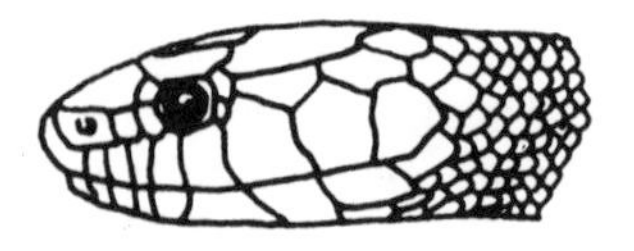

Fig. 20. Head of northern ringneck snake, 3× natural size

Size: From about 4 inches at hatching to nearly 24 inches. Most adults are between 10 and 16 inches long.
Variation: Two subspecies are recognized in Virginia. *Plate 22*
Diadophis punctatus edwardsi (Merrem). Northern Ringneck Snake. Neck ring usually complete. Belly usually unmarked.
Diadophis punctatus punctatus (Linnaeus). Southern Ringneck Snake. Neck ring usually incomplete. Belly with row of dark, half-moon-shaped spots.

Mitchell (1974) stated that the northern ringneck snake was found west of an imaginary line stretching from Prince George to Mecklenburg counties and that the southern ringneck snake was found in areas east of the line. The areas of intergradation, however, are not well known. Populations in the east-central area of the state often show a mixture of traits. Thus, individual snakes cannot properly be assigned to either subspecies without an extensive study of the variation in the local population.
Similar Species: See Table E above. Other snakes with ringlike neck markings either have keeled scales or a different dorsal coloration.

Habitat. Ringneck snakes prefer moist wooded areas. Both pine and hardwoods provide a suitable habitat if there is an abundance of rotting logs, old stumps, and loose bark to serve as hiding places. These snakes also live in cutover lands, sawdust piles, field edges, and occasionally suburban backyards.

Ringneck snakes are probably the most common snakes in Shenandoah National Park. One was found at the top of Stony Man Mountain (over 4,000 feet elevation) on the edge of the cliff. Others are commonly seen in the evening crawling across the boulders in the campsites or even in the rustic stone cottages. These snakes are also abundant in the Great Dismal Swamp, especially under the litter around old hunting camps. Dozens may be found in such areas along with skinks (*Eumeces* spp.) that share the habitat.

Range. Ringneck snakes are found from Nova Scotia and Ontario south to southern Florida and the Gulf Coast. Various subspecies are found as far west as California and south into Mexico. Ringneck snakes occur throughout the state of Virginia.

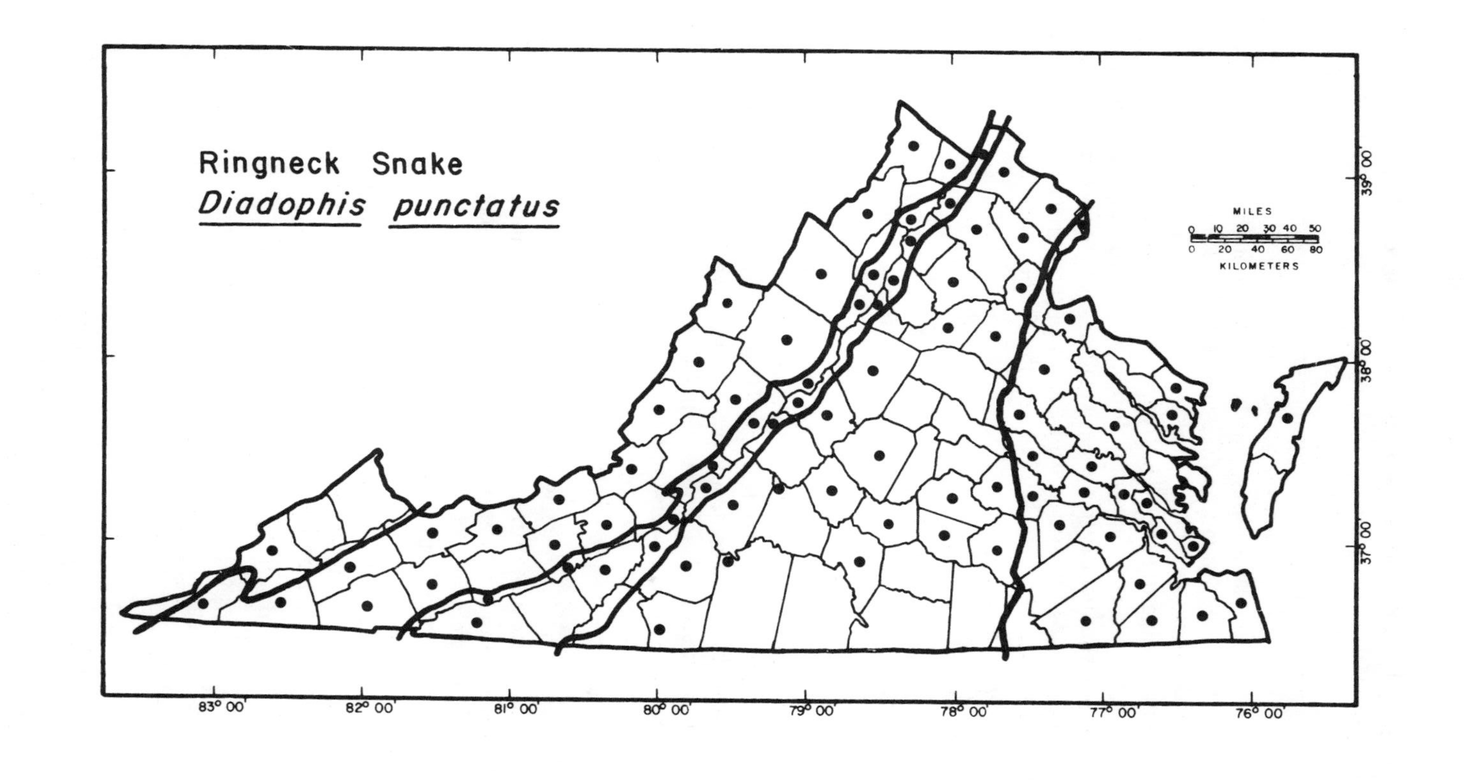
Ringneck Snake
Diadophis punctatus
MILES
0 10 20 30 40 50
0 20 40 60 80
KILOMETERS
39° 00'
38° 00'
37° 00'
83° 00'
82° 00'
81° 00'
80° 00'
79° 00'
78° 00'
77° 00'
76° 00'

Habits. Ringneck snakes display an interesting combination of habits in the various sections of their North American range. Some of the subspecies, including the southern ringneck snake in Florida, will twist and raise their tails (like a corkscrew) when approached by certain predators. This reveals the bright red ventral coloration and can be quite startling. Individuals of some subspecies, but apparently not those in Virginia, will feign death on occasion as a defensive reaction. Ringneck snakes are also among the most sociable of snakes. More often than not, two or more are found hiding together. This is apparently more than just the result of seeking similar environmental conditions. Two ringneck snakes were found on a summer day in the Big Meadows area of Shenandoah National Park under the same rock with an adult eastern garter snake and a large eastern milk snake. This kind of sociability could provide a meal for the milk snake.

Reproduction. From 2 to 10 elongate, whitish eggs are usually laid in rotting logs in June or early July. The 3½- to 5-inch young hatch during late summer. Several clutches of eggs may occasionally be laid together in a communal nest.

Food. Ringneck snakes are active little predators that feed on a variety of small animals including insects, earthworms, small snakes, small lizards, salamanders, and frogs. Uhler, Cottam, and Clarke (1939) examined five snakes from the George Washington National Forest. Food items, by volume, included salamanders 80%, ants 15%, and miscellaneous insects and arthropods 5%. These snakes usually swallow their prey alive but may use partial constriction in certain cases.

Enemies. The "corkscrew defense" mentioned above (which may or may not be used by Virginia specimens) is probably used against birds, since many other predators lack red vision and the display is directed upward. Kingsnakes and other serpent-eating snakes probably would ignore it if presented, although there is evidence that they are not colorblind.

Generally, ringneck snakes are subject to the same predation that other small snakes endure. Raccoons, opossums, skunks, black bears, shrews, and toads prey on these snakes.

In Captivity. Ringneck snakes are not the best captives, but they will survive in a large woodland terrarium if natural foods are available. Beware of keeping smaller snakes with them unless these snakes are intended to serve as food.

Folklore. Many country folks believe that ringneck snakes are the young of the eastern kingsnake. They supposedly add rings as they grow older until they achieve the complete adult pattern. Actually, young kingsnakes have the complete chain pattern at birth.

Location of Specimens. AMNH, ANSP, CAS, CM, CU, CWM, DU, FMNH, GMU, GMW, JHU, LC, MCZ, MLBS, NVCC, ODU, PNSC, SDSNH, UK, UMMZ, USNM, VCU, VMNH, VPI and SU.

WORM SNAKES
(Genus *Carphophis*)

At first glance, worm snakes with their brown and pink cylindrical bodies are often mistaken for earthworms. Closer inspection, however, reveals tiny eyes, scales, a forked tongue, and their other reptilian traits. The dorsal scales are smooth, and the anal plate is divided.

Worm Snake
(*Carphophis amoenus*)

Other Common Names: Ground snake, blind snake. *Plate 23*

Description. Adult: Plain brown or dark purplish black above with an iridescent sheen. Belly and lower sides pink with the pink color extending upward onto the first several rows of dorsal scales. Head is small and pointed but is not distinct from the neck. Eyes are small. Body is cylindrical and moderately stout. Tail is moderately short and has a spinelike tip.
Juvenile: Juveniles are darker than adults.
Scalation: Dorsal scales smooth; 13 scale rows; anal plate divided. Loreal scale present; preocular scale absent (fig. 21).

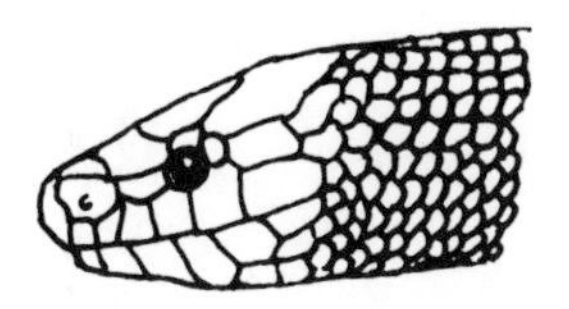

Fig. 21. Head of eastern worm snake, 3× natural size

Size: From about 3½ inches at hatching to about 13 inches. Most adults are between 7 and 11 inches long.
Variation: *Carphophis amoenus amoenus* (Say), the eastern worm snake, is the only subspecies occurring in the state. It intergrades, however, with the tan-colored midwest worm snake (*Carphophis amoenus helenae*) in western Virginia (Burger, 1975). Tan specimens form a small percentage of populations well into central Virginia.
Similar Species: Could be confused with the northern brown snake, the northern red-belly snake, and especially the earth snakes. (See Table F above.)

Habitat. Worm snakes live underground and under cover in pine woods, hardwood forests, old fields, pastures, and backyards. They are also par-

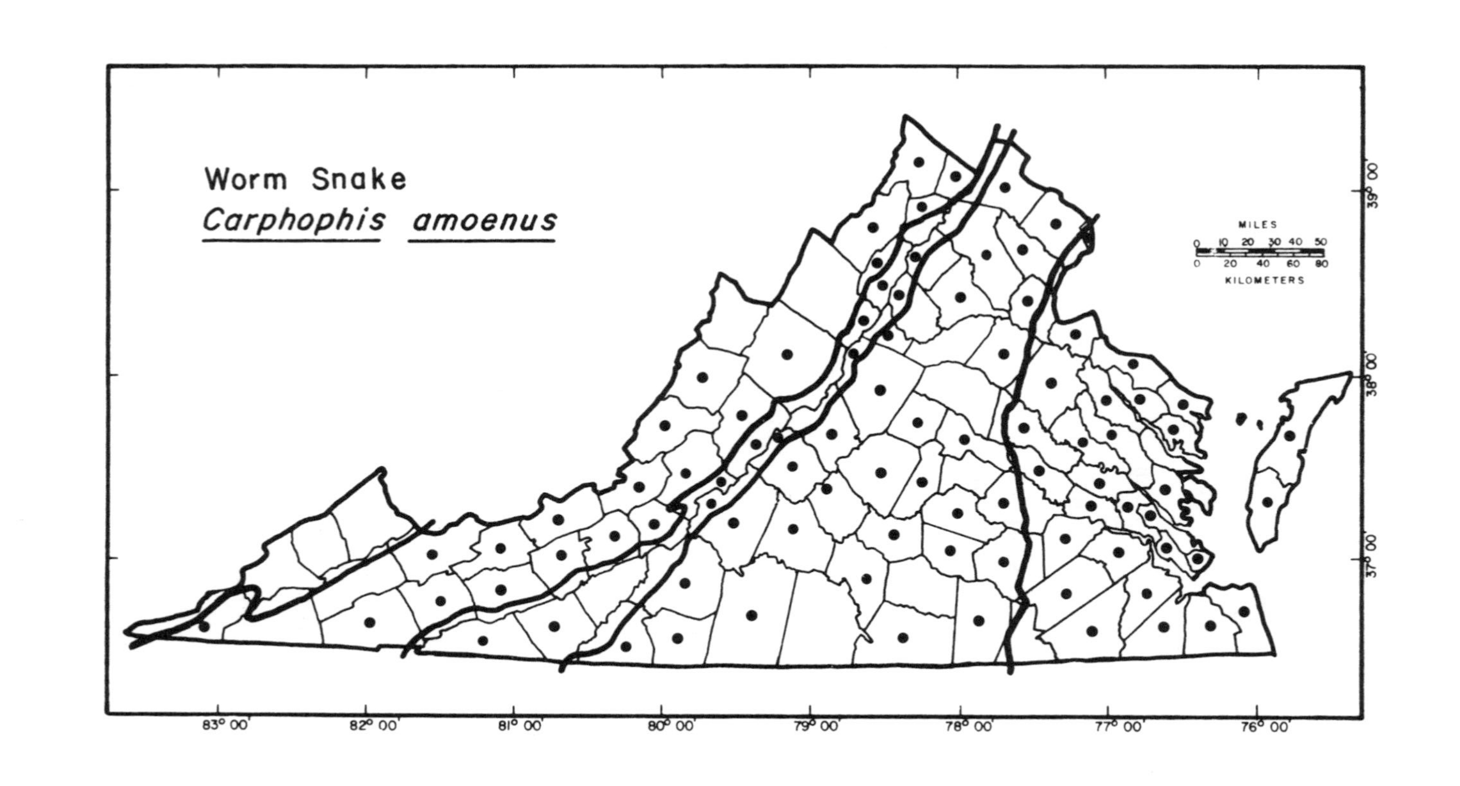

Worm Snake
Carphophis amoenus
MILES
0 10 20 30 40 50
0 20 40 60 80
KILOMETERS
39° 00'
38° 00'
37° 00'
83° 00'
82° 00'
81° 00'
80° 00'
79° 00'
78° 00'
77° 00'
76° 00'

tial to sawdust piles and rotten stumps. Their cylindrical, smooth-scaled body is ideal for traveling through the small passageways in the soil. They are often turned up by gardeners, farmers, and others who work the earth. Small boys often discover these little snakes when looking for crawly playthings under boards or logs.

Range. The range of this species extends from Massachusetts, New York, Pennsylvania, Ohio, Indiana, Iowa, and Nebraska south to central Georgia, southern Alabama, and Louisiana. The range extends westward into central Oklahoma and Kansas. This species occurs throughout Virginia. Burger (1975) recorded a worm snake from Lee County that was apparently an intergrade between the eastern worm snake (*Carphophis amoenus amoenus*) and the midland worm snake (*Carphophis amoenus helenae*).

Habits. Worm snakes, like other small species of snakes, are quite docile when handled. They do not strike or bite in self-defense but instead attempt a squirming escape. They probe between fingers with both the shovellike head and the sharp-tipped tail. A secretive species, these snakes are seldom found in the open, except perhaps at night or after heavy rains.

Barbour, Harvey, and Hardin (1969) found that worm snakes in Kentucky travel over a limited area in the course of their normal activities, returning periodically to certain areas. Home range sizes varied from 23 m^2 to 726 m^2 with an average size of 253 m^2.

Reproduction. From 2 to 8 (usually 5) eggs are laid during June and July in decaying logs and stumps, sawdust piles, and other similar areas. The size and shape of the eggs are quite variable. Smaller clutches contain larger, more elongated eggs that may be over an inch in length. Larger clutches contain eggs that are half that length. The young, which hatch in late summer, are 3 to 4 inches long. One clutch of 11 eggs is either a record number or an example of communal nesting.

Food. Worm snakes eat earthworms, soft-bodied insects, small slugs, and snails. Uhler, Cottam, and Clarke (1939) examined the stomachs of four worm snakes taken in the George Washington National Forest. Two of them had fed exclusively on earthworms and two on fly larvae. All prey is usually swallowed alive.

Enemies. Secretive habits protect worm snakes from a variety of potential predators. The sharp-tipped tail may serve as a defensive weapon against soft-skinned predators such as ambystomid salamanders that occasionally attack small snakes. When on the surface of the ground, worm snakes have fallen prey to such diverse creatures as toads and opossums. A specimen in the collection of the Carnegie Museum was recovered from the stomach of a mole kingsnake (*Lampropeltis calligaster*).

In Captivity. Worm snakes do well enough in a woodland terrarium but are seldom seen. Since observation of the creatures is a major reason for keeping snakes, the problem can be somewhat overcome by using an "ant farm" type of structure to house worm snakes.

Location of Specimens. AMNH, ANSP, CHM, CM, CWM, DS, DU, DWL, ESU, FMNH, GMU, GMW, JHU, LC, MCZ, MLBS, MNHS, NVCC, PNSC, SDSNH, UI, UMMZ, USNM, VCU, VIMS, VMNH, VPI and SU.

MUD AND RAINBOW SNAKES
(Genus *Farancia*)

These are among the most beautiful and unusual snakes found in Virginia. Both are fairly large, red and black serpents of our eastern lowlands. They have stiff tail tips (sometimes described as spines) that are normally used for maneuvering prey into better positions for swallowing and, to some extent, in defense. Most of the dorsal scales are smooth, and the anal plate is usually divided.

Despite the many similarities between these two species, they had long been placed in separate genera. Recently, however, the rainbow snake's genus, *Abastor,* was "absorbed" into the genus *Farancia,* the generic title under which the mud snake has long lived. In any case, the two snakes are closely related.

Mud Snake
(*Farancia abacura*)

Plate 24

Other Common Names: horn snake, hoop snake, red-bellied swamp snake.

Description. Adult: Shiny black and smooth above with the black extending in bars across the belly. The basic belly color is pink or red. This coloration extends up along the sides as bars or triangles. The body is relatively stout and muscular, and the head is barely distinct from the neck. Tail tip is stiff and blunt.
Juvenile: Similar to adult, but reddish side markings may join across the anterior portion of the back to form complete bands. Tail tip is sharp.
Scalation: Dorsal scales mostly smooth except above the anal region where they are keeled; 19 scale rows; anal plate is usually divided but occasionally single. Loreal scale present; preocular scale absent (fig. 22).

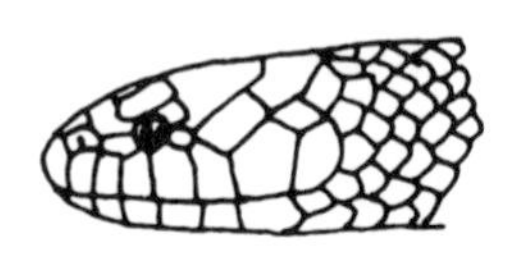

Fig. 22. Head of eastern mud snake, natural size

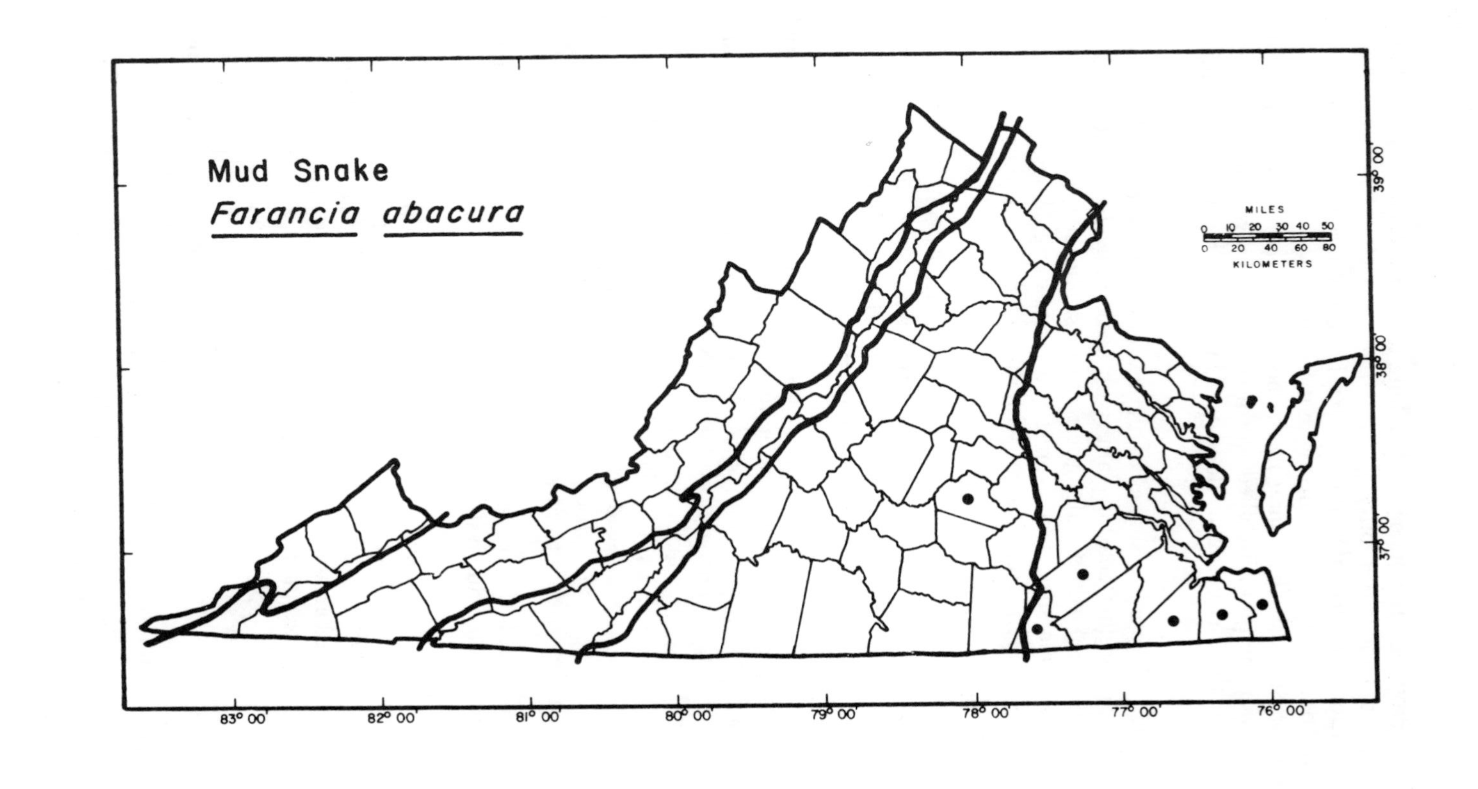
Mud Snake
Farancia abacura
MILES
0 10 20 30 40 50
0 20 40 60 80
KILOMETERS
39° 00'
38° 00'
37° 00'
83° 00'
82° 00'
81° 00'
80° 00'
79° 00'
78° 00'
77° 00'
76° 00'

Size: From about 7 inches at hatching to over 6½ feet. Most adults are between 3 and 5 feet long.
Variation: *Farancia abacura abacura* (Holbrook), the eastern mud snake, is the only subspecies occurring in the state.
Similar Species: The young might possibly be confused with red-belly snakes or the pinkish-bellied worm snakes. However, the bright ventral color of these two snakes does not extend up the sides as triangles or bars.

Habitat. These are snakes of the swamps and other lowlands of the southern Virginia coastal plain, with some individuals extending their range into the Piedmont along certain rivers. Mud snakes have been found in brackish marshes, dark cypress and gum swamps, bogs, and other wet areas. They are partial to areas with thick stands of various broad-leaved water plants.

Range. Mud snakes are found on the coastal plain from Virginia to southern Florida and west into eastern Texas. They range northward in the Mississippi Valley to Missouri, Kentucky, and Illinois. Mud snakes are found only in the southeastern portion of Virginia.

Habits. Mud snakes are seldom seen. Not only is their habitat avoided by most people, but they are quite secretive. Much time is spent burrowing in mud, muck, and sand. They are among the most aquatic of Virginia's snakes and are, of course, excellent swimmers. Occasionally mud snakes are seen in the open, especially at night or during rains. When captured, this species, like the rainbow snake, rarely bites. Instead, the snake tries to hide its head under its coil and will stab at its attacker with its blunt-tipped tail. Most predators will grab the tail, thus giving the snake an opportunity to twist itself free.

Reproduction. Up to 100 or more (record 104) eggs are laid during the summer. Average clutch size is between 25 and 50. The 7- to 9-inch young hatch in September.

Most snakes do not protect or care for their eggs once they have been deposited in a suitable site. A number of instances have been reported, however, where both captive and wild female mud snakes have remained coiled around their eggs from the time they were deposited until they hatched except for short periods to feed, molt, or defecate (Meade, 1940; Riemer, 1957). This behavior is apparently unique among North American snakes. Brooding Indian and diamond pythons actually aid the incubation of their eggs by increasing their body temperature as much as 5.5° to 7.3°F above the temperature of the surrounding environment. This temperature increase is brought about through spasmodic body contractions or shivering (Harlow and Grigg, 1984). No such increase in body temperature has been noted in the mud snake.

Longevity. Maximum known age: 18 years, 10 days (Bowler, 1977).

Food. Amphiumas and sirens, large eellike salamanders, are the favorite prey. The mud snake uses its tail tip in positioning the salamander for swallowing. Other prey includes tadpoles, frogs, small salamanders, and fish.

Enemies. It can be assumed that, under stable conditions, snakes and other wild animals have evolved the ability to produce enough young to maintain their populations. Mud snakes lay large numbers of eggs (more than any other Virginia species) to counter a large natural mortality. Predation by larger animals accounts for a significant portion of this destruction, and, as with the rainbow snake, the predation is probably particularly heavy among the eggs and young.

In Captivity. Feeding is often a problem in keeping a mud snake. A supply of amphiumas is usually difficult to obtain. Some mud snakes will accept a strip of meat which has been rubbed with a frozen amphiuma. The scent stimulates the snake's feeding response. Mud snakes can be kept in a semiaquatic terrarium.

Folklore. The "hoop snake" tale is usually applied to the mud snake and to the rainbow snake. Supposedly, the snake takes its tail into its mouth, forms a hoop, and rolls after the nearest human. It then tries to "sting" the person with its tail. Should the snake jab a tree instead, the poor plant immediately wilts and dies. The whole story is, of course, nonsense.

Location of Specimens. ANSP, CM, DS, GMW, HSH, USNM, VCU, VPI and SU.

Rainbow Snake
(*Farancia erytrogramma*)

Plate 25

Other Common Names: mud snake, hoop snake, horn snake, red swamp snake, sand snake, sand hog.

Description. Adult: A shiny, iridescent, and very beautiful snake with three red stripes on a bluish-black background. A broad reddish stripe bordered with black spots runs down the center of the belly and is flanked by yellow-orange coloration. The body is relatively stout, and the head is barely distinct from the neck. The tail is short and ends in a spinelike tip.
Juvenile: Similar to adult.
Scalation: Dorsal scales smooth except above the anal region where they may be keeled; 19 scale rows; anal plate is usually divided in most regions but is often single in Virginia specimens (Richmond, 1954). Loreal scale present; preocular scale absent (fig. 23).

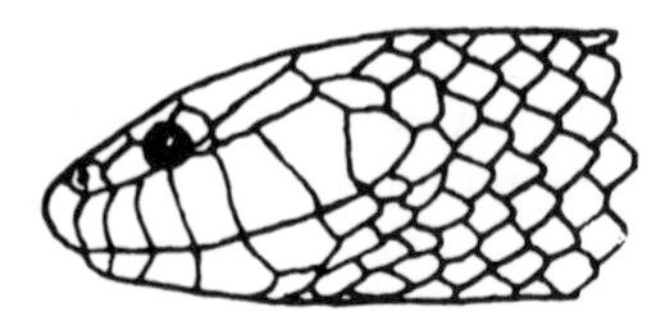

Fig. 23. Head of rainbow snake, natural size

Size: From 8 inches or more at hatching to more than 60 inches. Adults are usually between 36 and 48 inches in length.
Variation: *Farancia erytrogramma erytrogramma* (Latreille), the rainbow snake, is the only subspecies occurring in the state.
Similar Species: None.

Habitat. Swamps, marshes (freshwater and brackish), or slow-moving streams and adjacent sandy soils generally under 100 feet in elevation are the key environmental conditions for this species. Its habitat generally includes suitable muck and sand for burrowing and eels for food. Many areas on Virginia's coastal plain provide such conditions. This snake is probably more common in these areas than is generally realized, but it is rarely seen in the open.

Range. The range of this snake is restricted to the southeastern United States. It is found on the coastal plain from southern Maryland to south-central Florida and westward to Louisiana. In Virginia, rainbow snakes are found only in the eastern portions of the state.

Habits. Rainbow snakes can be characterized as both aquatic and burrowers. They are excellent swimmers but more often slowly prowl along the stream or swamp bottoms. At other times they burrow into muck or mud. They have been found in dry sand at depths of as much as 10 feet. More commonly they are turned up at plow depth in fields near swampy streams or marshes. Richmond (1945) reported as many as 20 rainbow snakes having been plowed out of a 10-acre field in one day in New Kent County. Young snakes have been found beneath boards, logs, and other ground debris. When captured, rainbow snakes seldom bite. However, they do use their sharp-tipped tail to probe at a collector's arms and hands.

Richmond (1945) noted that these snakes apparently have no definite period of hibernation, since active specimens have been recorded during every month of the year. A hawk has been observed eating a rainbow snake on the ice of a marsh in February.

Reproduction. Up to 52 eggs are laid in an underground cavity in sandy soil, usually during July. The clutch of leathery white eggs is deposited from 4 to 18 inches below the surface of the ground in open, exposed, dry, sandy fields. The 8- to 11-inch young hatch in the fall, probably over-

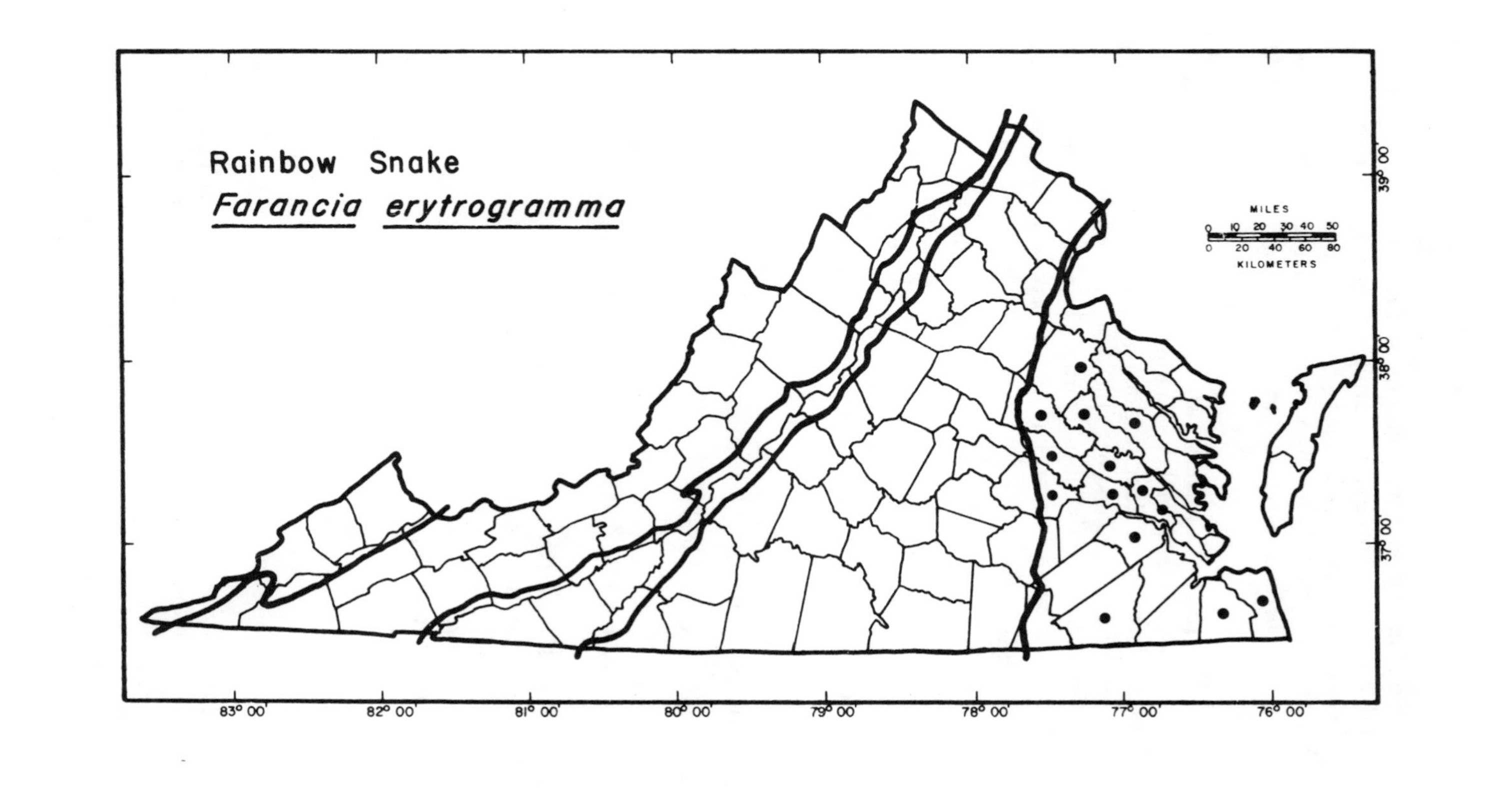
Rainbow Snake
Farancia erytrogramma
MILES
0 10 20 30 40 50
0 20 40 60 80
KILOMETERS
39° 00'
38° 00'
37° 00'
83° 00'
82° 00'
81° 00'
80° 00'
79° 00'
78° 00'
77° 00'
76° 00'

winter on land by burrowing into the soil near the nest, and move overland to an aquatic area in early spring. Males may reach sexual maturity by the end of their second, or the beginning of their third growing season, while females probably reach maturity by the end of the third year and possibly by the end of the second year (Gibbons, Coker, and Murphy, 1977).

Food. Eels (*Anguilla rostrata*) are the principal food of adult rainbow snakes. Young rainbow snakes eat tadpoles and small frogs in addition to eels and salamanders. After catching an eel, the rainbow snake usually climbs out of the water and onto the shore or into the exposed roots of a bald-cypress tree or a streamside shrub. The prey is swallowed head first. Swallowing is rapid at first, but then proceeds more slowly. Thus, the snake often rests with the fish's tail dangling from its jaws (Neill, 1964).

Enemies. The eggs and newly hatched young are probably much more susceptible to predation than adults. The nest locations are often in the same areas as those chosen by various aquatic turtles. These areas are in turn a mecca to egg-loving predators such as raccoons, opossums, skunks, and kingsnakes. Richmond (1945) recorded several instances of predation on rainbow snakes by hawks in Virginia.

In Captivity. Rainbow snakes can be kept in semiaquatic terrariums. However, they are difficult to keep unless eels are available for feeding.

Folklore. See mud snake.

Location of Specimens. ASU, CM, CU, GMW, HSH, LSUMZ, MSWB, UC, UFMNH, UMMZ, USNM, VCU, VMNH, VPI and SU.

RACERS
(Genus *Coluber*)

Racers are medium to large snakes with long, slender bodies. They have smooth dorsal scales and a divided anal plate.

Black Racer
(*Coluber constrictor*)

Other Common Names: black snake, racer, blue racer. *Plates 26–27*

Description. Adult: Black with no pattern. Gloss is dull like a gun barrel rather than shiny. Chin and throat are white, but the remainder of the underside is dark bluish-gray. Auffenberg (1955) studied the morphological variations in eastern forms of this species.
Juvenile: Dark grayish or brownish blotches on a light gray background. Belly is gray with small dark spots. The eyes are very large. Distinct pat-

tern until after the first year. Darkening occurs rapidly; 24-inch specimens are almost pure black.
Scalation: Dorsal scales smooth; 17 scale rows; anal plate divided. Loreal and preocular scales present (fig. 24).

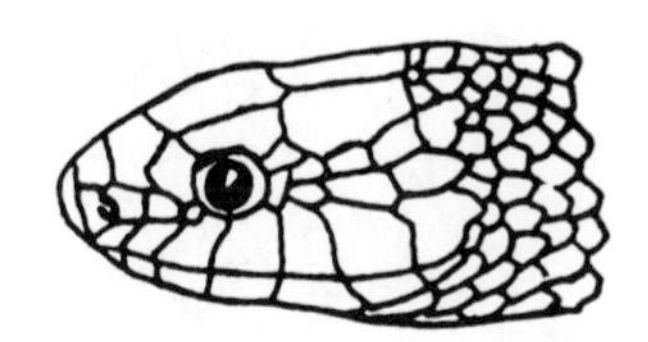

Fig. 24. Head of northern black racer, natural size

Size: From about 10 inches at hatching to over 6 feet. Most adults are generally between 3 and 5 feet long.
Variation: *Coluber constrictor constrictor* Linnaeus, the northern black racer, is the only subspecies occurring in the state.
Similar Species: Most apt to be confused with the black rat snake (see Table C above). Juveniles may be mistaken for other small blotched snakes (see Table D above).

Habitat. Black racers are quite adaptable. Habitats include weedy fields, cultivated field edges, brushy areas, briar patches, cutover areas, pine woods, and young hardwoods. These snakes are found from rocky mountain hillsides to live oak thickets near the Atlantic Ocean. Neill (1958) recorded this species as occurring in association with brackish water. Dry areas, however, are preferred to wet.

Range. Black racers are found from Maine, New York, and Ohio south to southern Florida and westward to eastern Texas and eastern Oklahoma. They occur throughout Virginia.

Habits. Black racers are diurnal and basically terrestrial, although they climb shrubs and small trees in search of food or when pursued. Despite their speed, racers will quickly stand and fight if they feel cornered. The defensive display is impressive. The head is held high and the mouth is held partially open as the snake constantly shifts its position and strikes. As with many species, the tail tip shakes nervously and will produce a rattling sound if among dry leaves. When actually attacked, the snake may hide its head under its coiled body.

Reproduction. Female black racers usually deposit 12 to 18 eggs in June. Some clutches, however, may contain as many as 40 eggs. Hatching occurs in late July or August with hatchlings averaging about 12 inches in length. Juvenile black racers undergo a developmental (ontogenetic)

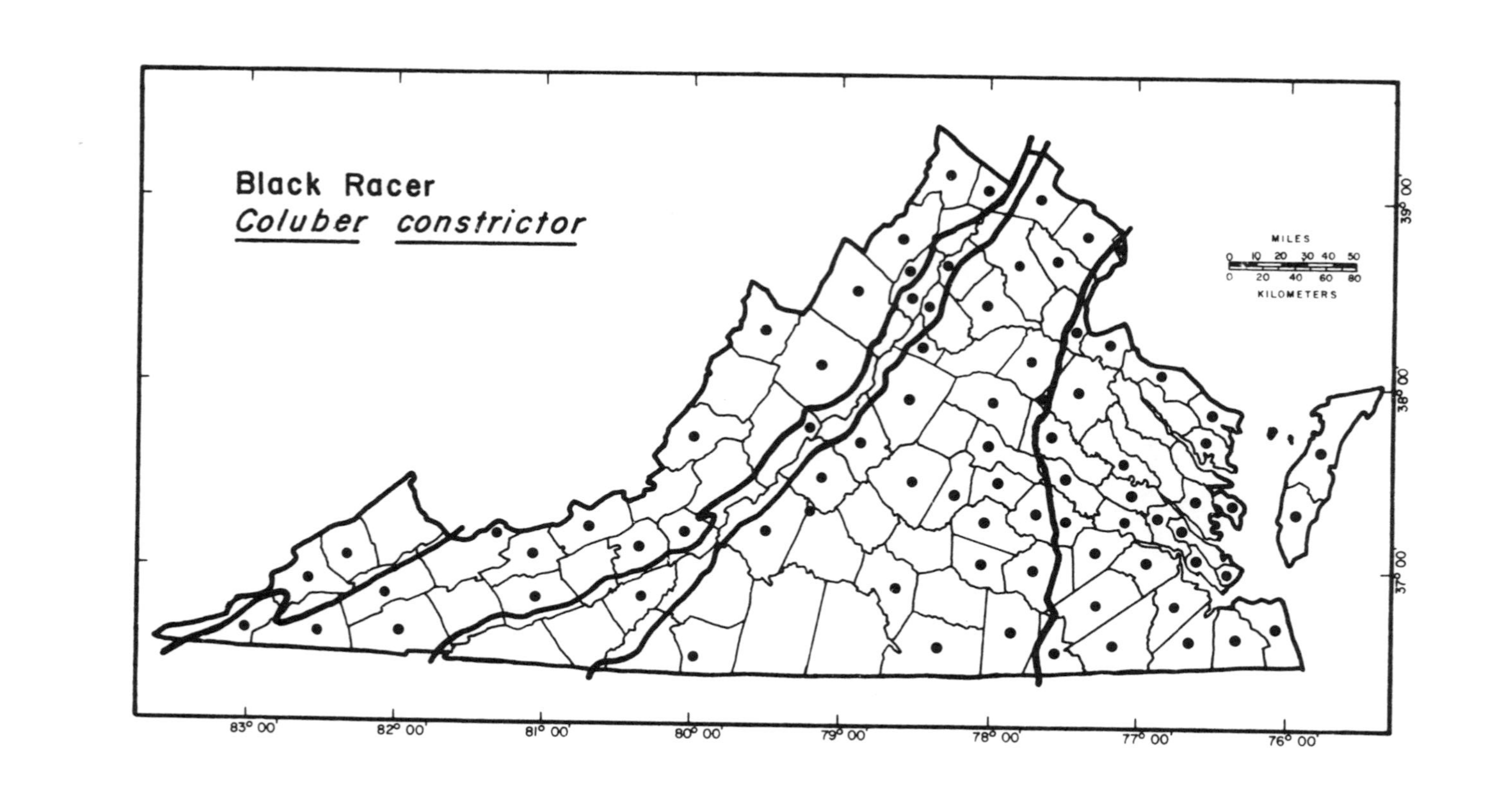
Black Racer
Coluber constrictor
MILES
0 10 20 30 40 50
0 20 40 60 80
KILOMETERS
39° 00'
38° 00'
37° 00'
83° 00'
82° 00'
81° 00'
80° 00'
79° 00'
78° 00'
77° 00'
76° 00'

change as they grow older. Young snakes are gray and brown with distinct blotches but gradually turn black during their second or third year. The blotches seem to melt into dead leaves and weeds and undoubtedly are of substantial survival value to the small snakes. The adults, being very active on defense and in escape, probably find more value in their energy-absorbing black coloration.

Food. Black racers eat a wide variety of animal life including small mammals, birds and their eggs, snakes (including the young of poisonous species), lizards, frogs, toads, and some insects, especially grasshoppers. Uhler, Cottam, and Clarke (1939) examined the food contents of 16 black racer stomachs from the George Washington National Forest. Major food items, by volume, included snakes 26%, birds 18%, shrews 12%, caterpillars and moths 10%, frogs 9%, moles 6%, lizards 6%, chipmunks and squirrels 5%, and other insects and arthropods 5%. Hoffman (1945) recorded a Virginia specimen which had fed on an adult dusky salamander (*Desmognathus fuscus*). Richmond and Goin (1938) noted one specimen from New Kent County that had its stomach full of June bugs. These racers are not constrictors, despite their scientific name. They usually swallow their prey alive, but may throw a coil over a struggling victim to pin it down.

Enemies. Black racers, along with black rat snakes, are widely recognized as harmless and beneficial and thus are not usually persecuted by man. Speed and vigorous defense serve as protection from some natural predators. Raccoons, foxes, opossums, hawks, and kingsnakes take their share. Lewis (1940) recorded an adult male opossum consuming a 3-foot 8-inch black racer in Amelia County.

In Captivity. Black racers generally are poor captives. Always ready to bite, they remain nervous and rarely can be tamed. Although ready feeders, most specimens are not very hardy.

Folklore. The speed of black racers is often exaggerated. Believed by many to outrun horses, they in fact can be run down by a man in open land. Maximum speed is less than 4 MPH (Oliver, 1955). Another tale claims that hanging a dead black snake on a fence post will bring rain. Eventually rain will fall, of course, and this somehow justifies the belief. A common superstition among children is that "if you change directions quickly when a black snake chases you, the snake will break its back." There may be some truth to the story that black racers chase people. In at least one case known to Clifford, a black racer has attacked without apparent provocation. However, they do not chase pregnant women particularly, as some believe.

Location of Specimens. AMNH, ANSP, CFR, CM, CU, CWM, DU, ESU, GMU, GMW, LC, LSUMZ, MCZ, NVCC, PNSC, TAMU, UI, UM, UMMZ, USFWS, USNM, VCU, VIMS, VMNH, VPI and SU, YPM.

GREEN SNAKES
(Genus *Opheodrys*)

Green snakes are slender, attractive snakes—as green as the vegetation in which they live. They are related to the racers and whip-snakes, with whom they share such characteristics as large eyes, quick movements and nervous dispositions. However, they are far more docile when captured. Camouflage, rather than viciousness, is a green snake's defense.

Two species of green snakes occur in North America and both are found in Virginia. They are the smooth green snake (smooth dorsal scales) and the rough green snake (keeled dorsal scales). Both species have divided anal plates.

Rough Green Snake
(*Opheodrys aestivus*)

Plate 28

Other Common Names: garden snake, grass snake, vine snake, keeled green snake.

Description. Adult: Plain light green above and white, cream, or yellow underneath, often with a greenish cast. Body is very slender with the head being wider than the neck. On rare occasions, greenish-brown individuals are found. Preserved specimens turn blue.
Juvenile: Similar to adult.
Scalation: Dorsal scales strongly keeled; 17 scale rows; anal plate divided. Loreal and preocular scales present (fig. 25).
Size: From 7 inches at hatching to more than 40 inches. Usual adult size is 18 to 30 inches.
Variation: Two subspecies are recognized in Virginia.

Plate 28

Opheodrys aestivus aestivus (Linnaeus). Rough Green Snake. Occurs statewide except on the barrier islands.
Opheodrys aestivus conanti Grobman. Barrier Island Rough Green Snake. Occurs only on the Virginia barrier islands.
Similar Species: Could only be mistaken for the eastern smooth green snake which has smooth dorsal scales in 15 rows at mid-body.

Fig. 25. Head of rough green snake, 2× natural size

Habitat. The rough green snake is an inhabitant of areas of thick, green vegetation. Small trees, bushes, briar patches, and tangles of vines are favorite areas. These snakes are often attracted to the lush vegetation overhanging streams and lakes. Home gardeners are occasionally startled by a green snake gliding through the runners or cruising through the squash. They should be welcomed to a garden, however, for hornworms and other destructive caterpillars are favorite green snake foods. This species is one of several Virginia snakes that can maintain its population in suburban developments and even city parks if shrubs and hedges are available. The rough green snake has also been recorded in association with brackish water habitats (Neill, 1958).

Range. This species ranges from New Jersey, Delaware, Maryland, Ohio, Indiana, Illinois, and Missouri south to the Florida Keys and west into Texas, Oklahoma, and Kansas. The rough green snake is found throughout Virginia although records for the western mountainous region of the state are rare.

Habits. The rough green snake is distinctly arboreal, although Richmond (1952) recorded several instances of this snake in the water in New Kent County. It hunts, hides, and rests in vegetation several feet above the ground. This snake is quite graceful and exhibits impressive body control. It can stretch over half its body length straight out in midair when reaching for a branch. A startled rough green snake will often pull back its head as if to strike but, instead of biting, it suddenly darts away. This is a diurnal species which often spends the entire night sleeping in the same vegetation it prowled during daylight. Occasionally, it may be knocked from its sleeping perch, as was one along a Dismal Swamp canal which dropped into a canoe full of Boy Scouts. Luckily, instead of jumping from the boat, the scouts turned on a flashlight and found, to their relief, an innocent rough green snake.

Reproduction. Up to a dozen eggs are laid in rotting trees, logs, or stumps during June and July. The 7- to 8-inch young hatch in late summer.

Food. Mainly grasshoppers, crickets, caterpillars, and spiders are swallowed alive. Small frogs are also sometimes eaten. Uhler, Cottam, and Clarke (1939) examined two specimens from the George Washington National Forest. Food items, by volume, were caterpillars 55%, snails and slugs 20%, grasshoppers 15%, and harvestmen 10%.

Enemies. Green snakes are recognized as harmless by most people who see them. Seeing one, however, is no easy matter. It must be difficult as well for predators that hunt by sight (and color) such as hawks. Those that locate prey more by scent such as opossums, skunks and kingsnakes probably are more important predators. Plummer (1990) recorded predation by kingsnakes, black racers, and bluejays in Arkansas. Large spiders have been known to kill young green snakes.

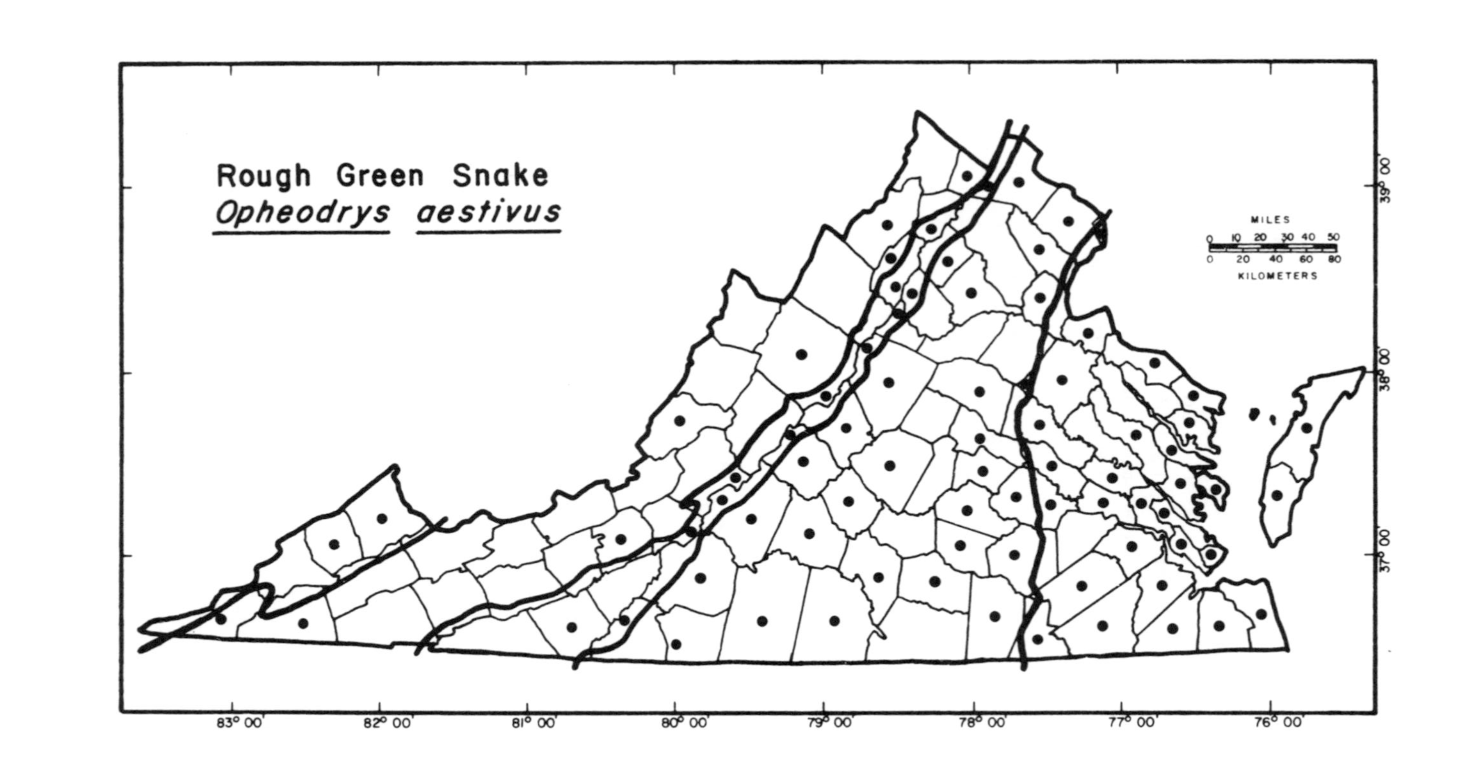
Rough Green Snake
Opheodrys aestivus
MILES
0 10 20 30 40 50
0 20 40 60 80
KILOMETERS
39° 00'
38° 00'
37° 00'
83° 00'
82° 00'
81° 00'
80° 00'
79° 00'
78° 00'
77° 00'
76° 00'

In Captivity. Normally a poor captive, this species usually remains nervous and refuses to eat. Smaller specimens may survive in a large terrarium supplied with vines, other plants, and slabs of bark to hide under. Providing live food is difficult in cold months unless crickets or mealworms are raised.

Folklore. The "glass snake" myth is sometimes mistakenly applied to the green snake. Should one be chopped in half, the snake supposedly pulls itself back together. This tale is more properly assigned to the eastern slender glass lizard (a legless lizard) which does have the ability to eventually regenerate a new tail, although it certainly cannot fuse together sections of its body.

Location of Specimens. AMNH, CFR, CM, CWM, DS, DWL, GMU, GMW, JHU, LC, ODU, OU, PNSC, SDSNH, TAMU, UMMZ, USNM, VCU, VIMS, VMNH, VPI and SU.

Smooth Green Snake (*Opheodrys vernalis*)

Other Common Names: grass snake, garden snake. *Plate 29*

Description. Adult: Bright green above and greenish-white, cream, or light yellow below. Body is slender. Preserved specimens turn blue.
Juvenile: Similar to adult but darker and less bright.
Scalation: Dorsal scales smooth; 15 scale rows; anal plate divided. Loreal and preocular scales present (fig. 26).

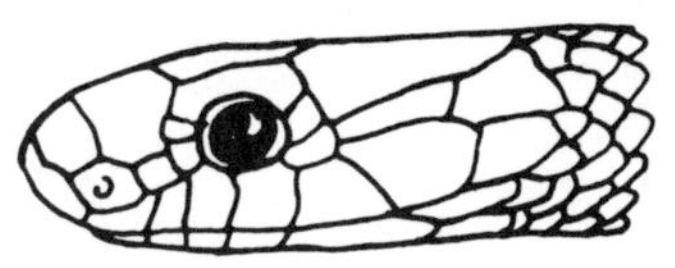

Fig. 26. Head of eastern smooth green snake, 3× natural size

Size: From about 5 inches at hatching to slightly over 24 inches. Adults are normally between 12 and 20 inches long.
Variation: *Opheodrys vernalis vernalis* (Harlan), the eastern smooth green snake, is the only subspecies occurring in the state.
Similar Species: Only the rough green snake is solid green above. It has keeled dorsal scales and 17 scale rows and is more elongated.

Habitat. Meadows and grassy fields are the usual haunts of this snake. It is also seen near bogs and open woods and in bramble patches. Sometimes it takes cover in rock piles and under logs.

Range. The smooth green snake is much more northern in its distribution than the rough green snake. Its main range extends from Nova Scotia

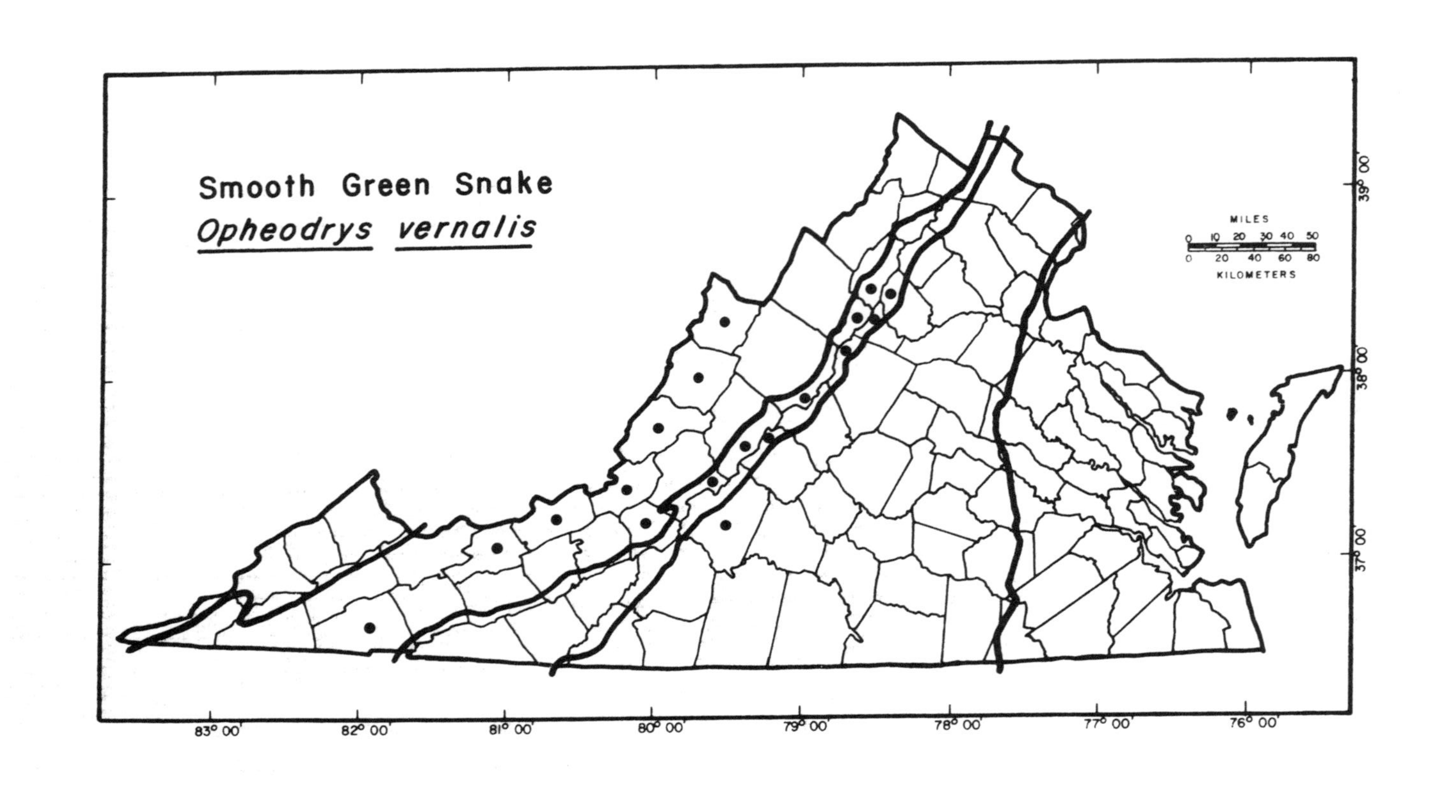
Smooth Green Snake
Opheodrys vernalis
MILES
0 10 20 30 40 50
0 20 40 60 80
KILOMETERS
39° 00'
38° 00'
37° 00'
83° 00'
82° 00'
81° 00'
80° 00'
79° 00'
78° 00'
77° 00'
76° 00'

and southern Canada south to Virginia and Missouri. The range extends westward into Nebraska, North Dakota, and Montana. A number of disjunct populations exist in various parts of the United States and Mexico. In Virginia, the smooth green snake is primarily restricted to the northern and central mountain regions.

Habits. Far more terrestrial than its keeled-scale relative, the smooth green snake is usually found at grass level rather than in shrubbery. It apparently tends to be more arboreal in wet areas, however. When surprised in the open, it glides quickly into nearby grass and becomes practically invisible. This species, like some others, will hibernate in large aggregations, at least in the north. Ant mounds seem to be favorite sites. Sociable habits also apparently carry over into warm weather since communal nesting sites have been found.

Reproduction. Female smooth green snakes deposit from 3 to 12 eggs during the summer. The 5-inch young often hatch less than a month after the eggs are laid.

Food. Insects and spiders make up the bulk of the diet. Salamanders, snails, slugs, centipedes, and millipedes are also eaten. Uhler, Cottam, and Clarke (1939) examined the stomach contents of five snakes taken in the George Washington National Forest. Major food items, by volume, were caterpillars 37%, spiders 32%, grasshoppers 20%, ants 10%, and snails and slugs 1%. Most prey is swallowed alive.

Enemies. Similar to those mentioned for the rough green snake.

In Captivity. This species is generally easier to maintain than the rough green snake. A large terrarium with a variety of living plants and hiding places is most suitable.

Folklore. See rough green snake account.

Location of Specimens. AMNH, CM, GMU, OSU, UMMZ, USNM, VPI and SU.

RAT SNAKES
(Genus *Elaphe*)

Rat snakes are large, powerful constrictors that kill their prey by wrapping their muscular bodies tightly around it. These snakes have both smooth and weakly keeled dorsal scales and a divided anal plate.

Rat snakes differ from other Virginia snakes in that they are flat-bottomed with the sides at almost a 90° angle to the ventral surface. Thus, in cross section the snakes are shaped like a loaf of bread rather than being oval. This body shape, in conjunction with powerful muscles and weakly keeled ventral scales, allows rat snakes to climb surprisingly smooth surfaces. The ventral keels provide sharp corners or projections that the snake can press against the bark of a tree, for example. Thus, these snakes can climb up the trunk of a large tree by wedging their bodies between the ridges of bark and hitching upward at least partly by means of the keeled ventral scales.

Corn Snake
(*Elaphe guttata*)

Plates 30–34

Other Common Names: red rat snake, pine snake, chicken snake.

Description. Adult: Bright reddish or brownish dorsal blotches, outlined in black, on a background that may be orange, gray, or light brown. Dorsal markings form a spear-point between the eyes. Belly is white with squarish black blotches.
Juvenile: Pattern similar to adult with dark reddish-brown blotches.
Scalation: Most dorsal scales very weakly keeled, lower rows smooth; usually 27 scale rows; anal plate divided. Loreal and preocular scales present (fig. 27).

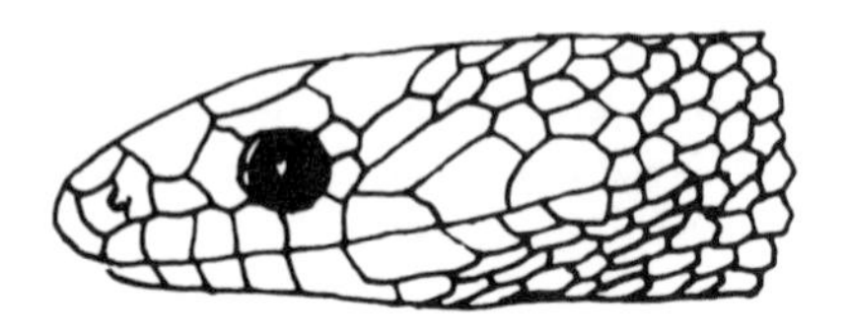

Fig. 27. Head of corn snake, 2× natural size

Size: From about 10 inches at hatching to a maximum of 6 feet. Most adults are between 2½ and 4 feet long.
Variation: *Elaphe guttata guttata* (Linnaeus), the corn snake, is the only subspecies occurring in the state.
Similar Species: See Table D above. The eastern milk snake and the mole snake are the most similar. Juvenile black rat snakes might be confused with young corn snakes. The spearpoint between the eyes is diagnostic.

Habitat. Corn snakes receive their name from a common prowling area—the edge of corn fields. They are also partial to other cultivated fields, open woodlots, rocky hillsides, cutover areas, and pine forests. These snakes are commonly found around barns and other outbuildings.

Range. The corn snake ranges from New Jersey south to southern Florida and westward into eastern Texas and Arkansas. Disjunct populations occur in Kentucky. This snake has been found throughout all but extreme southeastern Virginia.

Habits. Corn snakes are much more secretive than their closest Virginia relative, the black rat snake. They frequent mammal burrows and the pas-

Plate 1. Brown water snake (Nerodia taxispilota)

Plate 2. Red-belly water snake (Nerodia erythrogaster erythrogaster)

Plate 3. Northern water snake (Nerodia sipedon sipedon)

Plate 4. Midland water snake (Nerodia sipedon pleuralis)

Plate 5. Queen snake (Regina septemvittata)

Plate 6. Belly pattern of queen snake (Regina septemvitata)

Plate 7. Glossy crayfish snake (Regina rigida)

Plate 8. Northern brown snake (Storeria dekayi dekayi)

Plate 9. Midland brown snake (Storeria dekayi wrightorum)

Plate 10. Northern red-belly snake (Storeria occipitomaculata occipitomaculata)

Plate 11. Eastern garter snake (Thamnophis sirtalis sirtalis)

Plate 12. Head of eastern garter snake showing extended tongue

Plate 13. Eastern garter snake consuming a fish

Plate 14. Eastern ribbon snake (Thamnophis sauritus sauritus)

Plate 15. Eastern smooth earth snake (Virginia valeriae valeriae)

Plate 16. Rough earth snake (Virginia striatula)

Plate 17. Eastern hognose snake (Heterodon platirhinos)

Plate 18. Eastern hognose snake bluffing

Plate 19. Head of eastern hognose snake showing upturned snout

Plate 20. Eastern hognose snake "playing dead"

Plate 22. Northern ringneck snake (Diadophis punctatus edwardsi)

Plate 21. Eastern hognose snake with eggs

Plate 23. Eastern worm snake (Carphophis amoenus amoenus)

Plate 24. Eastern mud snake (Farancia abacura abacura)

Plate 25. Rainbow snake (Farancia erytrogramma erytrogramma)

Plate 26. Northern black racer (Coluber constrictor constrictor)

Plate 27. Juvenile northern black racer (Coluber constrictor constrictor)

Plate 28. Rough green snake (Opheodrys aestivus)

Plate 29. Eastern smooth green snake (Opheodrys vernalis vernalis)

Plate 30. Corn snake (Elaphe guttata guttata)

Plate 31. Corn snake in coiled striking position

Plate 32. Corn snake depositing eggs

Plate 33. Corn snake shedding skin

Plate 34. Corn snake consuming a mouse

Plate 35. Adult black rat snake (Elaphe obsoleta obsoleta)

Plate 36. Juvenile black rat snake (Elaphe obsoleta obsoleta)

Plate 37. Northern pine snake (Pituophis melanoleucus melanoleucus)

Plate 38. Northern pine snake (Pituophis melanoleucus melanoleucus)

Plate 39. Eastern kingsnake (Lampropeltis getula getula)

Plate 40. Eastern kingsnake hatching

Plate 41. Black kingsnake (Lampropeltis getula niger)

Plate 42. Eastern milk snake (Lampropeltis triangulum triangulum)

Plate 43. "Coastal plain milk snake" (Lampropeltis triangulum triangulum x Lampropeltis triangulum elapsoides *intergrade)*

Plate 44. Scarlet kingsnake (Lampropeltis triangulum elapsoides)

Plate 45. Mole kingsnake (Lampropeltis calligaster rhombomaculata)

Plate 46. Northern scarlet snake (Cemophora coccinea coccinea)

Plate 47. Southeastern crowned snake (Tantilla coronata)

Plate 48. Northern copperhead (Agkistrodon contortrix mokasen)

Plate 49. Juvenile northern copperhead (Agkistrodon contortrix mokasen)

Plate 50. Eastern cottonmouth (Agkistrodon piscivorus piscivorus)

Plate 51. Timber rattlesnake (Crotalus horridus horridus)

Plate 52. Timber rattlesnake (Crotalus horridus horridus)

Plate 53. Canebrake rattlesnake (Crotalus horridus atricaudatus)

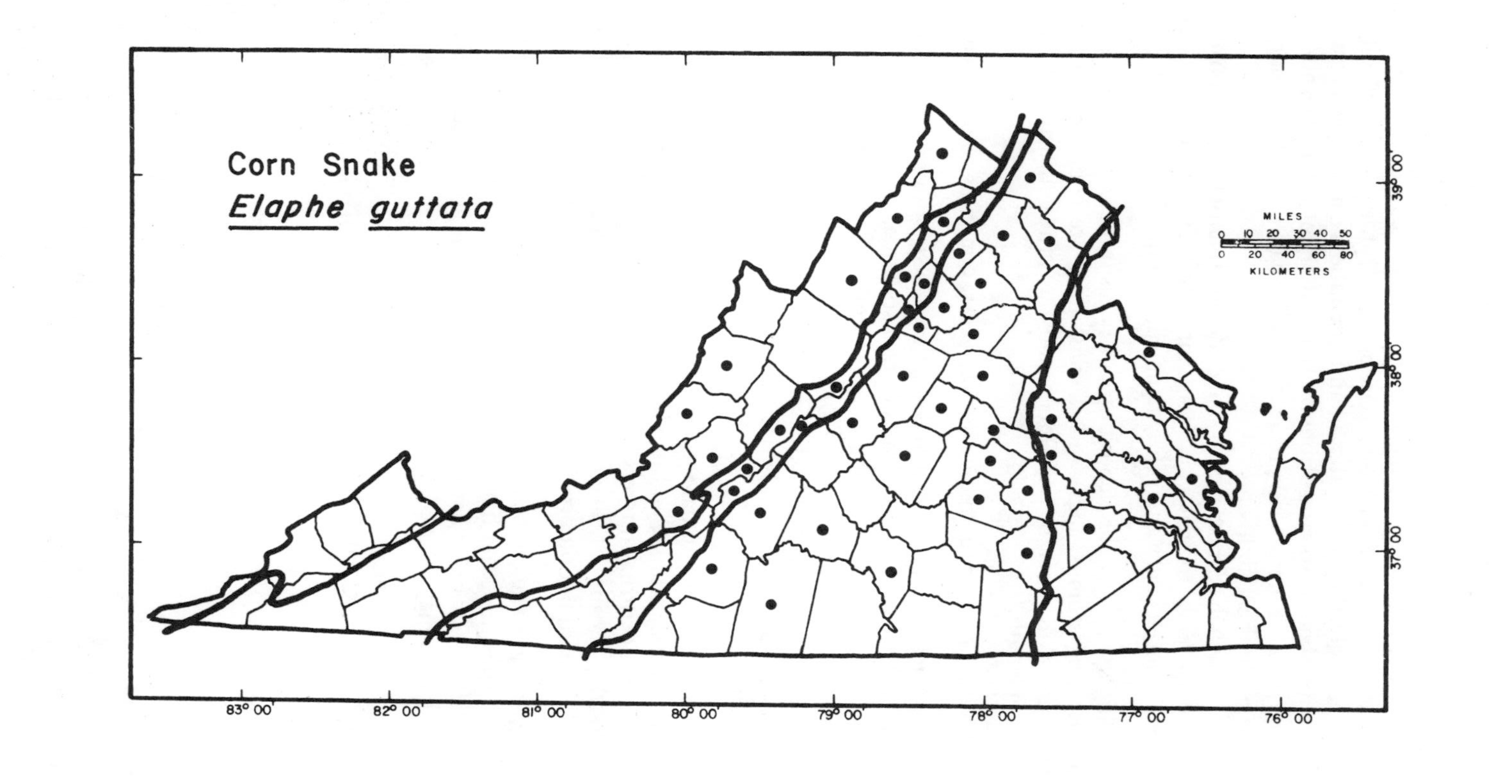
Corn Snake
Elaphe guttata
MILES
0 10 20 30 40 50
0 20 40 60 80
KILOMETERS
39° 00'
38° 00'
37° 00'
83° 00'
82° 00'
81° 00'
80° 00'
79° 00'
78° 00'
77° 00'
76° 00'

sageways in decayed root systems and rock crevices. Individuals are commonly found beneath boards, pieces of bark, rotting logs, and piles of leaves. At night they prowl abroad and often the most successful method of locating them is to cruise country roads after sundown. Despite their terrestrial and subterranean proclivities, corn snakes have the remarkable rat snake climbing ability.

When cornered, corn snakes defend themselves vigorously, rising in coils ready to strike and constantly shifting to face the attacker. However, once caught and gently handled, they quickly tame down.

Reproduction. Females deposit up to 24 eggs from early to middle summer. Rotting stumps, logs, and sawdust piles are common nest sites. Hatching takes place in September, and the young measure about 12 inches in length.

Longevity. Maximum known age: 21 years, 9 months (Bowler, 1977).

Food. Mice, small rats, and moles are favorite foods of this constrictor. Birds, bird eggs, bats, and shrews are also eaten. Uhler, Cottam, and Clarke (1939) examined two specimens from the George Washington National Forest. One contained a skink, while the other had fed upon a field mouse and a wood-boring beetle. Because of their preference for rodents, these snakes are welcomed by most farmers. Young corn snakes feed on lizards and frogs, especially skinks and treefrogs.

Enemies. Natural predators include foxes, bobcats, opossums, raccoons, weasels, skunks, and hawks. Corn snakes are probably more susceptible to nocturnal hunters like raccoons but safer from diurnal ones like hawks. Because of their secretive ways, corn snakes are not often attacked by man, but on occasion one may be killed as a "copperhead."

In Captivity. The corn snake is highly valued as a pet. Beautiful colors, gentle disposition, and hardiness make this species a favorite. Feeding is simple if mice or other natural food is available. If not, pieces of chicken or small eggs may be tried.

Location of Specimens. AMNH, ANSP, CFR, CM, CWM, ESU, GMW, LC, MCZ, NVCC, UI, USFWS, USNM, VCU, VMNH.

Black Rat Snake
(*Elaphe obsoleta*)

Plates 35–36

Other Common Names: black snake, pilot black snake, chicken snake, mountain black snake.

Description. Adult: Shiny black above with light specks on the edges of, or between, some of the scales. The specks are usually white but may be reddish. The anterior one-third of the belly is yellowish-white with scat-

tered black squares; farther back the belly becomes gray and is unmarked. Some snakes from the Dismal Swamp region are olive or dark tan with four dark stripes. These show the influence of the southern coastal subspecies, the yellow rat snake, *Elaphe obsoleta quadrivittata.* A 47-inch albino black rat snake was taken in Rockbridge County in 1949 (Carroll, 1950). Another albino individual was taken in Westmoreland County in 1957 (Hensley, 1959).
Juvenile: Dark gray or brown blotches on a light gray background. Cream-colored belly with black squares. Dark stripe extends backward from eye to lower edge of upper labial scales. The rate of darkening varies with each individual and a trace of pattern is often carried throughout life. Generally, however, a 30-inch specimen is a *black* snake.
Scalation: Most dorsal scales weakly keeled, lower rows smooth; 25–27 scale rows; anal plate divided. Loreal and preocular scales present (fig. 28).

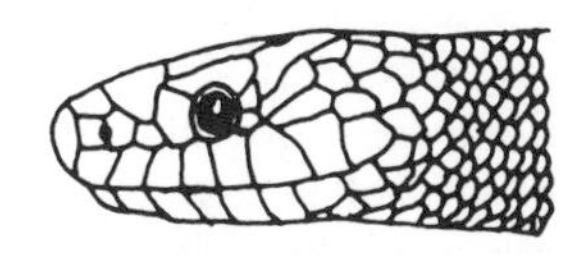

Fig. 28. Head of black rat snake, natural size

Size: From about 12 inches at hatching to over 8 feet. The black rat snake holds the record for being the longest snake ever officially recorded in the United States. Conant and Collins (1991) gives the maximum length as 101 inches, although Wright and Wright (1957) give the maximum length as 108 inches. The usual adult size is 3½ to 5½ feet. Many farmers who protect their black rat snakes have 6- to 7-foot specimens dwelling in their barns.
Variation: *Elaphe obsoleta obsoleta* (Say), the black rat snake, is the only subspecies occurring in the state.
Similar Species: Adults are most likely to be mistaken for black racers, kingsnakes, and the black phase of the hognose snake (see Table C above). Juveniles could be mistaken for various blotched species (see Table D above).

Habitat. Black rat snakes are generally associated with moderate to large trees and/or buildings. These snakes are found in a variety of woodlands—extensive forests or small woodlots, usually hardwoods but sometimes conifers. They frequently take up residence in barns and other farm buildings (sometimes including the farmer's house!). Their shed skins are often found draped in outbuildings. Rural villages and even suburban areas can provide suitable habitats particularly if large hollow trees are available.

Although certainly not semiaquatic, black rat snakes do not avoid wooded watercourses and swamps. In the Dismal Swamp region, for instance, they may be seen in the trees on Lake Drummond's shore or overhanging the Northwest River.

Range. The black rat snake ranges from southern Ontario, Massachusetts, Vermont, New York, Michigan, Wisconsin, and southeastern Minnesota south to Georgia, Mississippi, and Louisiana. This species is found west as far as Oklahoma, Kansas, and Nebraska. It occurs throughout Virginia.

Habits. This species is one of the most arboreal of Virginia's snakes. Probably only the rough green snake and perhaps the brown water snake spend as much time in trees and vegetation above the ground surface. In pure climbing ability, it exceeds any species in the state. Black rat snakes are able to negotiate smooth-barked trees or the side of a wooden barn. Climbing enables the snakes to reach food—birds and squirrels in trees and rats and mice in barns. Black rat snakes are also at home on the ground. They may crawl along the woodland floor, prowl through brush piles, or explore the cavities of a fallen tree. They also have the often fatal habit of lying out on country roads in early morning or late afternoon to raise the body temperature. This species, like several others, shakes the tip of its tail when frightened and, when in dry leaves, produces a good imitation of the buzzing of a rattlesnake. Homing to a food source has been reported by Weatherhead and Robertson (1990) in Ontario.

Reproduction. Although as many as 24 eggs may be deposited in late June or July, the usual clutch size is 8 to 12 eggs. The eggs are usually laid in well-decayed wood with stumps, logs, and hollow trees being the usual sites. A clutch of 15 eggs was discovered in a sawdust pile in Fairfax County (Klimkiewicz, 1972). Several females may deposit eggs in the same location. Hatching occurs in September and early October. The young are between 11 and 16 inches long.

Longevity. Maximum known age: 20 years, 1 month, 23 days (Bowler, 1977).

Food. Adult black rat snakes feed on practically any warm-blooded prey they can kill by constriction and swallow. Rats, mice, birds, and bird eggs are the usual victims, but animals as large as squirrels and small rabbits have been eaten by 6-foot snakes. Instances have been recorded of these snakes feeding on big brown bats (Silver, 1928). Uhler, Cottam, and Clarke (1939) examined the stomach contents of 85 black rat snakes taken from the George Washington National Forest in Virginia. Major food items, by volume, included mice 32%, birds 31%, chipmunks and squirrels 15%, rabbits 9%, and shrews 4%. Young rat snakes feed to a large extent on treefrogs and other small frogs and lizards, all of which they eat alive. Small mice are killed by constriction, however. In captivity,

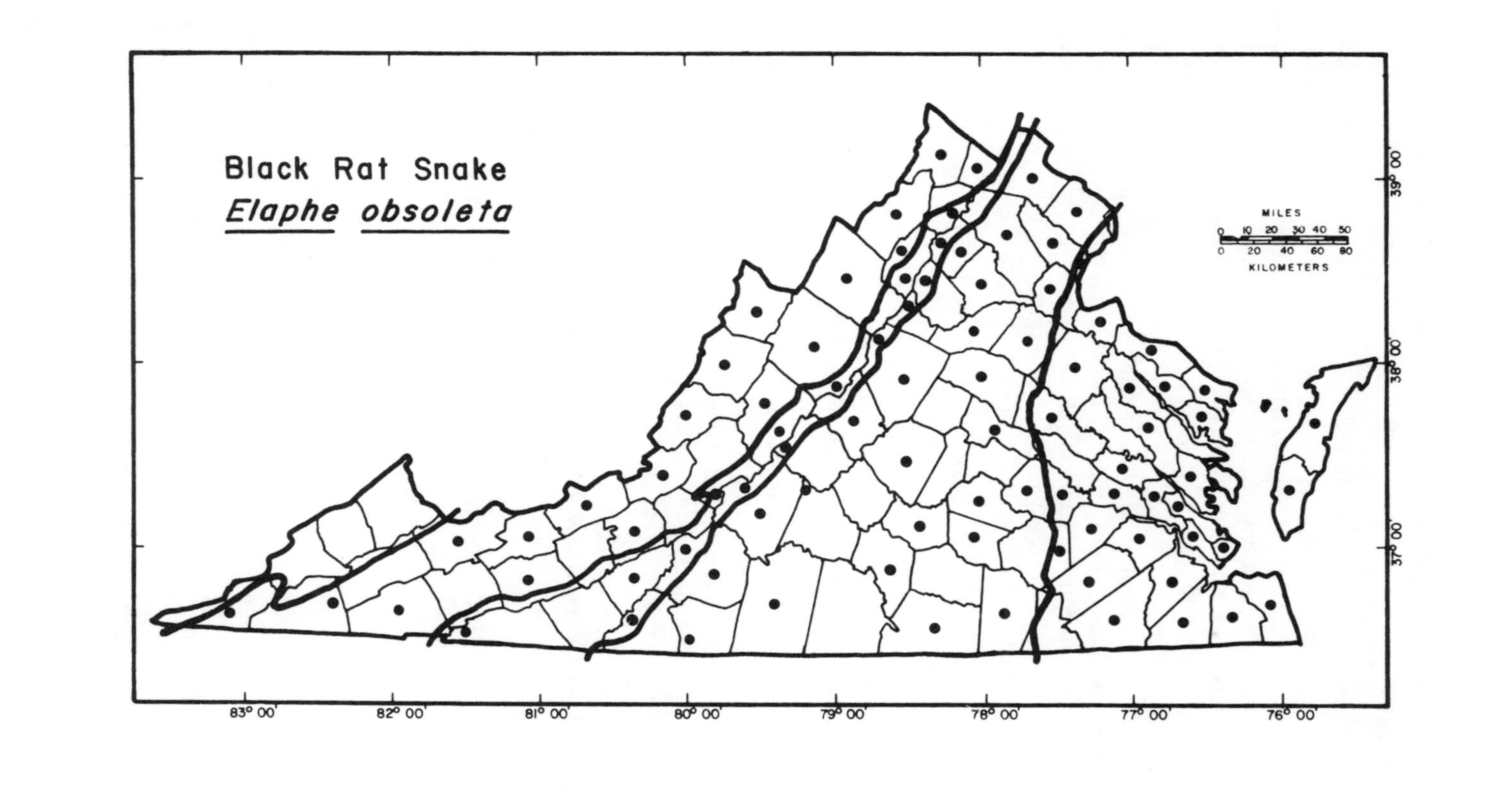
Black Rat Snake
Elaphe obsoleta
MILES
0 10 20 30 40 50
0 20 40 60 80
KILOMETERS
39° 00'
38° 00'
37° 00'
83° 00'
82° 00'
81° 00'
80° 00'
79° 00'
78° 00'
77° 00'
76° 00'

many specimens will eat such odd morsels as fried chicken and warm bologna.

Enemies. Because of their rodent-eating habits, in combination with their easy recognition, black rat snakes are freer from man's destruction than most other species. Most farmers protect their black rat snakes and may even bring in a few from the woods to release in their barns. In fact, some farmers are quite proud of the 6- or 7-foot specimens that have lived in their barns for years. The snakes sometimes do get into trouble in the hen house, however.

Natural predators include foxes, bobcats, opossums, raccoons, weasels, skunks, and hawks. As expected, smaller specimens are far more susceptible. Large black rat snakes have on several occasions kept Clifford's four beagles at bay, beagles that twice killed tough old tomcats that made the mistake of climbing into their pen.

In Captivity. Black rat snakes do very well in captivity. They eat readily and will often eat certain processed meats and store-bought eggs, which can be more convenient than natural foods. Although they vigorously defend themselves when first captured, most specimens tame quickly after some handling. The largest specimens are often the most docile.

Folklore. This species is known as the pilot black snake in certain mountain areas. According to folklore, they guide rattlesnakes to safety during times of danger; thus the name "pilot." Another tale, which is also applied to the black racer, claims that hanging a dead black snake on a fence post will bring rain.

Location of Specimens. AMNH, ANSP, CHM, CM, CU, CWM, DS, DU, ESU, GMU, GMW, JHU, LC, MLBS, MSWB, NVCC, ODU, PNSC, UK, UMMZ, USFWS, USNM, VCU, VIMS, VMNH, VPI and SU.

PINE SNAKES, BULLSNAKES, AND GOPHER SNAKES (Genus *Pituophis*)

The genus *Pituophis* contains one species that ranges from coast to coast. This species has been divided into six subspecies that are known as bull, gopher, and, in our area, pine snakes.

These snakes are large, powerful constrictors and, where common, can be valuable agents for rodent control. They have keeled dorsal scales and an undivided anal plate. An enlarged nose plate (rostral scale) and somewhat countersunken lower jaws are adaptations for burrowing.

An outstanding characteristic of these snakes is their ability to hiss loudly when disturbed. Their unique keel-shaped epiglottis functions to magnify sound. When a breath of air is exhaled forcibly from the lung, it blows across the keeled edge of the epiglottis, which increases the loudness of the exhalant hiss (Martin and Huey, 1971). The sound is similar to that produced when a person blows upon a card held edgewise in front of their lips. The epiglottal keel may also vibrate from side to side to produce the fluttering sound.

Northern Pine Snake (*Pituophis melanoleucus*)

Plates 37–38

Other Common Names: bull snake, black and white snake.

Description. Adult: Dark blotches on white or light gray ground color. Blotches are black, turning brown near tail. Head seems small for the large body. Belly is white with dark spots on the sides. Has enlarged rostral scale.
Juvenile: Similar to adult but ground color may be pinkish.
Scalation: Dorsal scales keeled; 29 scale rows; anal plate undivided. Loreal and preocular scales present (fig. 29).

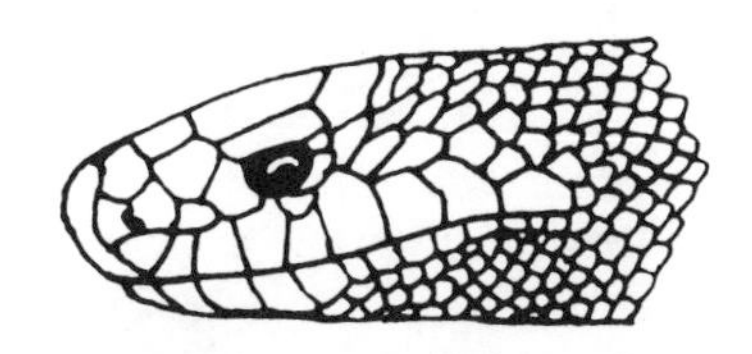

Fig. 29. Head of northern pine snake, natural size

Size: From about 15 inches at hatching to almost 7 feet. Usual adult size is between 4 and 6 feet.
Variation: *Pituophis melanoleucus melanoleucus* (Daudin), the northern pine snake, is the only subspecies occurring in the state.
Similar Species: Juveniles could be mistaken for young black rat snakes (see Table D above).

Habitat. In Virginia, northern pine snakes have thus far been found only in the mountains, where they inhabit dry ridges and hillsides. Vegetation includes scrub pine and thickets of laurel and rhododendron. In states to the north and south they inhabit coastal pine forests as well, but they are apparently absent from such regions in Virginia.

Range. The range of the northern pine snake is disjunct. One population is found in the pine barrens of the New Jersey coastal plain. The main segment of the range extends from western Virginia and Kentucky south through Tennessee to central Alabama and east through Georgia to the South Carolina coast. Northern pine snakes have thus far been found only in the western portion of Virginia. Dunn (1917) stated: "This snake is fairly well known in Virginia as the 'bull snake'. It does not seem to

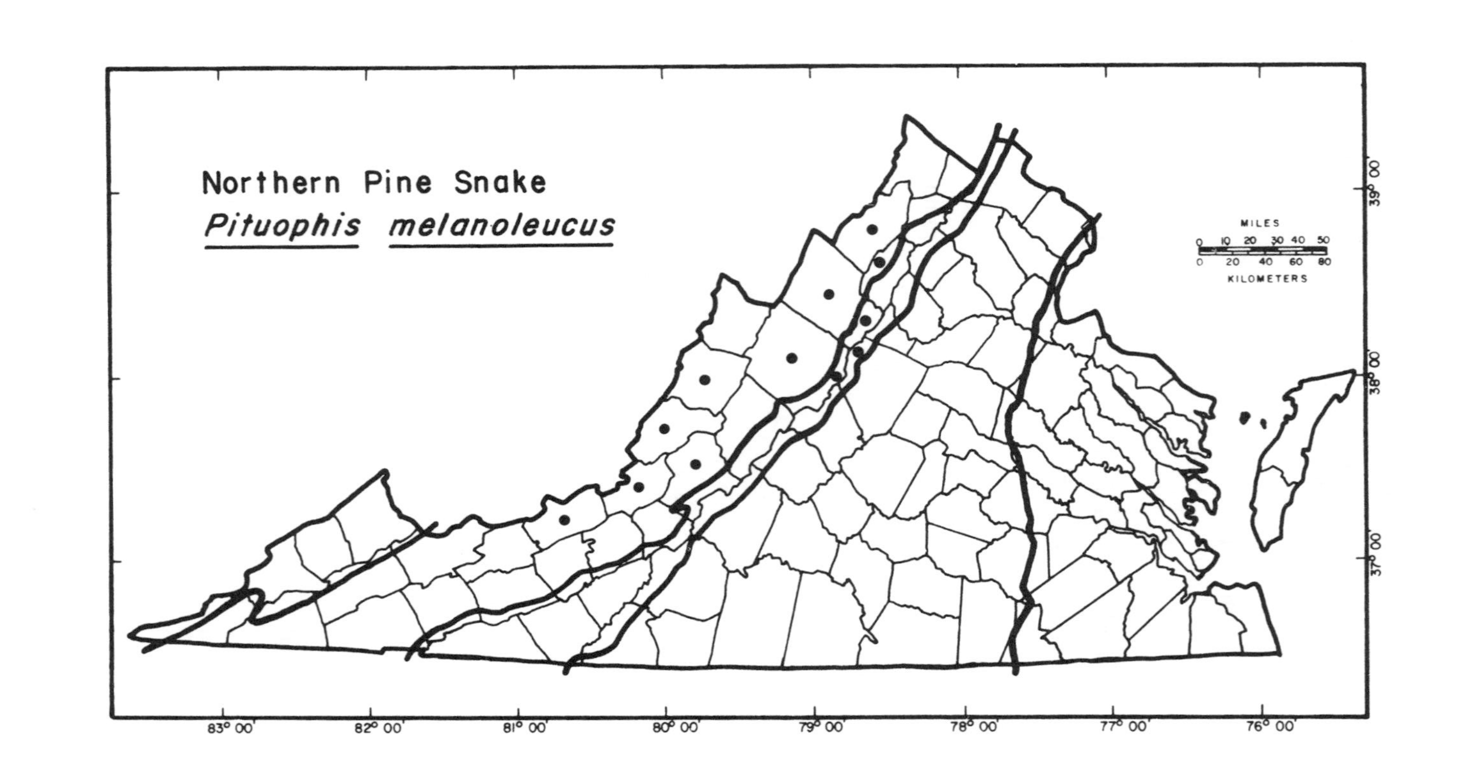
Northern Pine Snake
Pituophis melanoleucus
MILES
0 10 20 30 40 50
0 20 40 60 80
KILOMETERS
39° 00'
38° 00'
37° 00'
83° 00'
82° 00'
81° 00'
80° 00'
79° 00'
78° 00'
77° 00'
76° 00'

occur outside of the mountains as all of the many stories of this snake, reputed to reach a length of twelve feet, have their scene in the western tier of counties."

This species has officially been classified as "Status Undetermined" in Virginia (Mitchell, 1991).

Habits. Northern pine snakes are basically nocturnal. In addition, they spend much time underground and are so secretive that they may rarely be seen by local people. When cornered above ground they put on an impressive show: hissing loudly, vibrating the tail, and striking rapidly. Their prolonged hiss may be heard up to 50 feet away. However, they usually tame quickly after capture.

Reproduction. Up to 24 creamy-white, leathery-shelled eggs are laid in late spring and early summer. The eggs are usually deposited in a burrow several inches beneath the surface of the ground. Each egg is between 2 and 2½ inches long and between 1 and 1¼ inches in diameter. The 14- to 18-inch young hatch in August and September.

Longevity. Maximum known age: 20 years, 9 months, 2 days (Bowler, 1977).

Food. Rats, mice, moles, and other small mammals together with birds and bird eggs are eaten. Birds and small mammals are killed by constriction. When a large egg is eaten, it is swallowed for a short distance and then the strong neck muscles break the shell; thus, there is no loss of the liquid contents. The broken pieces of shell are either swallowed or disgorged.

Enemies. Because of their secretive habits and powerful build, adult northern pine snakes have few enemies. In one case, a northern pine snake held two grown raccoons at bay for an hour and then crawled away unmolested (Kauffeld, 1957). However, Burger et al. (1992) recorded subterranean predation in the hibernacula and nesting burrows by red foxes, striped skunks, and short-tailed shrews in New Jersey. These snakes have little contact with man and probably suffer most from habitat destruction.

In Captivity. Many individuals do well as captives; others remain too nervous. They are generally good feeders and are hardy.

Location of Specimens. HSH, USNM.

KINGSNAKES AND MILK SNAKES (Genus *Lampropeltis*)

Kingsnakes are medium- to large-sized terrestrial constrictors. They have smooth dorsal scales and an undivided anal plate.

Kingsnakes are well known for their habit of feeding on other snakes, both nonpoisonous and poisonous species. Although they do not seek out poisonous snakes, they will consume them if they are available. In this regard, it is interesting to note that kingsnakes are immune to the venom of our native poisonous snakes such as the copperhead, cottonmouth, rattlesnake, and coral snake.

Five varieties of this genus occur in Virginia: the common kingsnake (with two subspecies—the eastern kingsnake and the black kingsnake), milk snake (with two subspecies—eastern milk snake and scarlet kingsnake) and the mole kingsnake.

Common Kingsnake (*Lampropeltis getula*)

Plates 39–41

Other Common Names: chain snake, pine snake, thunder snake, black snake.

Description. Adult: Shiny black with a chainlike pattern of white, cream, or yellow rings or with scattered white or yellow specks arranged in a chainlike pattern. In rare cases, much of the white may be replaced by red. Belly is bluish-gray with white or yellowish squares or vice versa.
Juvenile: Black with distinct chainlike pattern.
Scalation: Dorsal scales smooth; usually 21 scale rows; anal plate undivided. Loreal and preocular scales present (fig. 30).

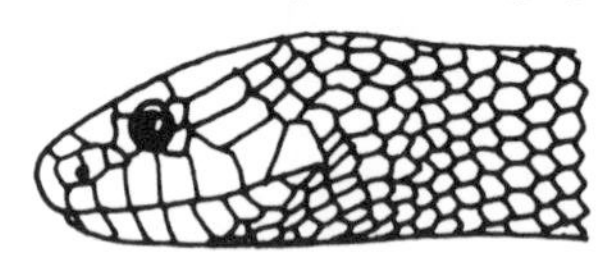

Fig. 30. Head of eastern kingsnake, natural size

Size: From about 9 inches at hatching to over 6 feet. Usual adult size is 3 to 4 feet.
Variation: Two subspecies are recognized in Virginia.

Plates 39–40

Lampropeltis getula getula (Linnaeus). Eastern Kingsnake. Black with chainlike pattern of white, cream, or yellow rings. Occurs statewide.

Plate 41

Lampropeltis getula niger (Yarrow). Black Kingsnake. Black with scattered white or yellow specks arranged in a chainlike pattern. Has only been recorded from Lee County.
Similar Species: Light flecks on black rat snakes do not form connected chainlike rings. (See Table C above)

Habitat. The eastern kingsnake has a distinct preference for moist areas: near swamps and marshes and along stream banks and lakeshores.

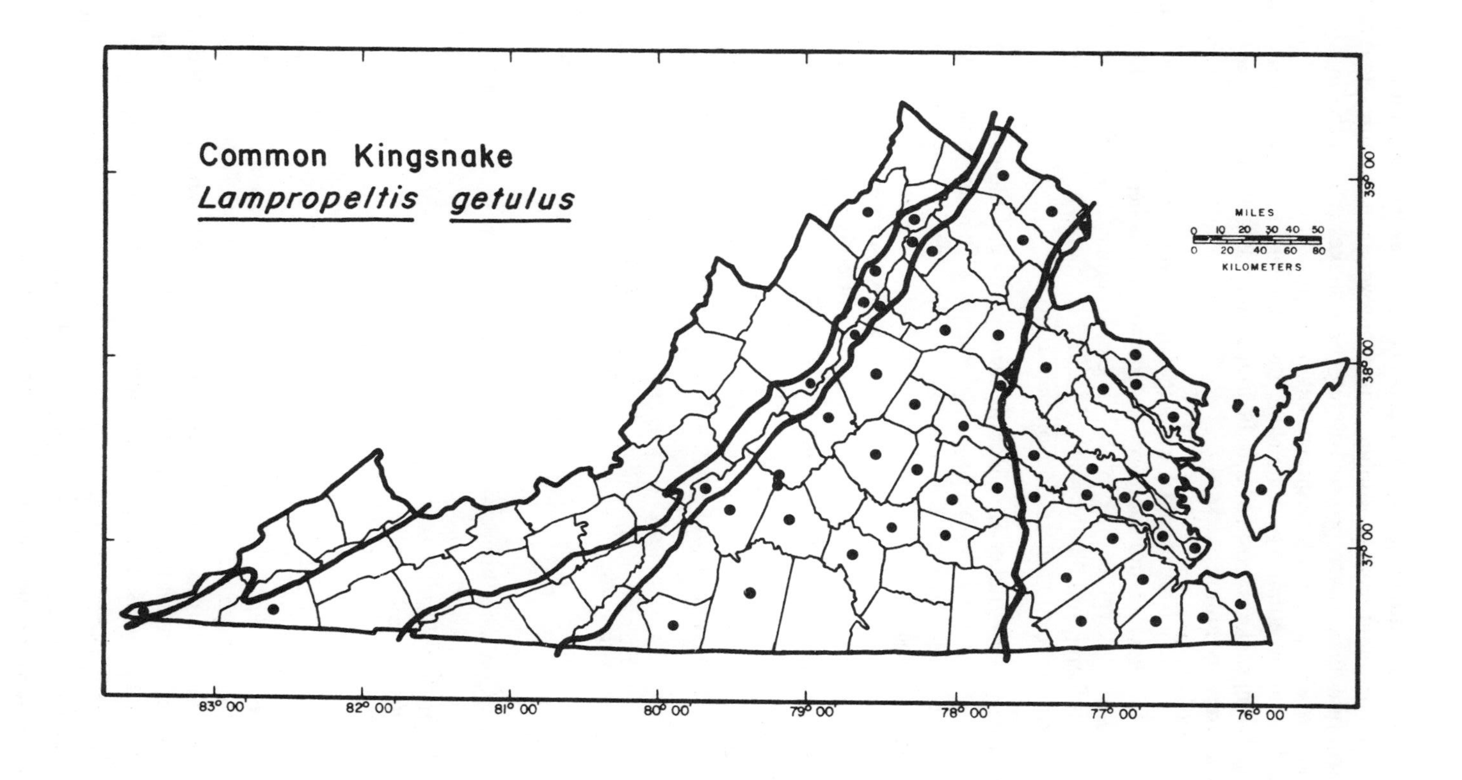
Common Kingsnake
Lampropeltis getulus
MILES
0 10 20 30 40 50
0 20 40 60 80
KILOMETERS
39° 00'
38° 00'
37° 00'
83° 00'
82° 00'
81° 00'
80° 00'
79° 00'
78° 00'
77° 00'
76° 00'

Its prey is abundant in such habitats. This species is also found in drier places including pine woods, field edges, cutover areas, pastures, and near farm buildings. Given proper habitat, these snakes can survive in suburban areas. One 5-foot specimen was found in a small marshy field that had been surrounded by a housing development for over 50 years. The black kingsnake is most likely to be found on rocky hillsides and pastures, near streams, and in thickets.

Range. The eastern kingsnake ranges from New Jersey and Maryland south to Alabama and south-central Florida. It has been found throughout most of Virginia. The black kingsnake ranges from West Virginia, Ohio, Indiana, and Illinois south to central Alabama. Its range enters only the extreme southwestern portion of Virginia.

Habits. These snakes are generally secretive by day, hiding under logs or boards or in rotten stumps or animal burrows. They are most often seen in the open at dusk or dawn or on cloudy days. A kingsnake seldom climbs. It is surprisingly fast on the ground, but when cornered it may coil, strike, and vibrate its tail. An alternative, or perhaps subsequent, defensive tactic is hiding its head under tight coils for protection. Once caught, kingsnakes are usually placid and rarely bite unless roughly treated.

Reproduction. Usually between 10 and 24 eggs are laid in late June or July. The 10- to 12-inch young hatch in August and September.

Longevity. Maximum known age:

Eastern Kingsnake—21 years, 5 months, 2 days (Bowler, 1977).

Black Kingsnake—13 years, 5 months, 17 days (Bowler, 1977).

Food. Common kingsnakes kill by constriction and are famous as snake-eaters. Garter and water snakes are common prey species. Other foods include lizards (especially skinks), reptile eggs, rodents, birds, and bird eggs. Juveniles will also eat small frogs and insects.

Kingsnakes, like other species, will sometimes devour enormous meals. A 3-foot eastern kingsnake once completely swallowed and digested a dead black rat snake that was only 3 inches shorter than itself.

Enemies. Generally recognized as beneficial, common kingsnakes do not suffer the persecution by man that many species receive. In addition, their strength and secretive habits give protection from many predators. Raccoons, opossums, skunks, and similar predators may occasionally attack kingsnakes, especially the young.

In Captivity. Gentle disposition, hardiness, and a lack of choosiness about food make this species a desirable captive. However, it is important to house kingsnakes apart from other snakes that are not meant to become a meal. When handled, a kingsnake will occasionally tie itself up in a ball of knots. At other times it may wrap several coils around an arm of the handler and display the powerful squeeze that helps it earn the name "king."

Folklore. The old superstition that mother snakes will swallow their young to protect them may have been partially caused by kingsnakes. An imaginative person might well interpret the sight of a kingsnake swallowing another snake as an act of parental love.

A superstition especially attached to this species is that should one be killed, a thunderstorm will surely follow. Another variation among children purports that the mere sight of a kingsnake will cause thunder.

Location of Specimens. AMNH, CHE, CM, CU, CWM, GMU, GMW, LC, MCZ, MSWB, NLU, NVCC, ODU, PNSC, SNP, UMMZ, USFWS, USNM, VCU, VMNH, VPI and SU.

Milk Snake and Scarlet Kingsnake (*Lampropeltis triangulum*)

Plates 42–44

Other Common Names: coral snake, red adder, house snake.

Description. This species consists of two subspecies which differ considerably in external appearance; hence they are treated separately.

Adult: Eastern Milk Snake: Ground color tan or gray. Three to five rows of reddish-brown, black-edged dorsal blotches. Center row of blotches largest and alternates with smaller, lateral blotches. Gray or tan Y- or V-shaped mark usually present at rear of head. Belly whitish with squarish black blotches giving checkerboard effect. An albino individual was taken near Blacksburg, Montgomery County, in 1955 (Hensley, 1959).

Adult: Scarlet Kingsnake: Reddish with yellow and black bands that completely encircle the body. Every other band is black; thus the sequence is red, black, yellow, black, red, black. Snout is reddish.
Note: East of the Blue Ridge, intergrade populations may exhibit coloration and pattern characteristics of both subspecies (see Plate 43).

Juvenile: Eastern Milk Snake: Similar to adult but with much redder blotches.

Juvenile: Scarlet Kingsnake: Similar to adult except that bands are red, black, and white (Groves and Sachs, 1973).
Scalation: Dorsal scales smooth; 19 scale rows; anal plate undivided. Loreal and preocular scales present (fig. 31).

Size: Eastern Milk Snake: From about 6 inches at birth to approximately 4 feet. Most adults are between 24 and 36 inches.

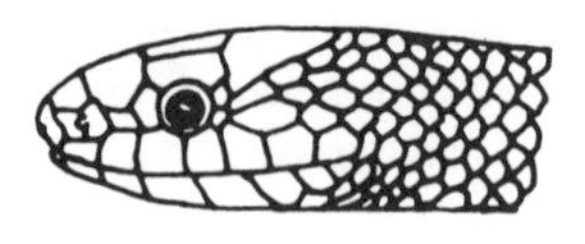

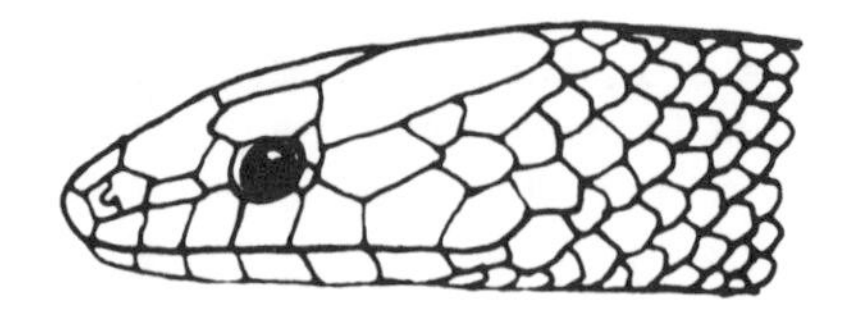

Fig. 31. a, *Head of eastern milk snake, natural size;* b, *head of scarlet kingsnake, 2× natural size.*

Size: Scarlet Kingsnake: From 6 to 8 inches at birth to approximately 24 inches. Most adults are between 15 and 20 inches.
Variation: Two subspecies are recognized in Virginia.

Plate 42

Lampropeltis triangulum triangulum (Lacépéde). Eastern Milk Snake. See description above. Occurs statewide.

Plate 43

Lampropeltis triangulum elapsoides (Holbrook). Scarlet Kingsnake. See description above. Occurs in southeastern Virginia.

Similar Species: Eastern Milk Snake: See Table D above.

Similar Species: Scarlet Kingsnake: Often confused with scarlet snake which lacks belly markings.

Habitat. Both of these snakes occupy a wide variety of habitats. They may be found on hillsides and in wooded areas, open fields, and stream and river floodplains.

Range: Eastern Milk Snake: This variety ranges from Ontario and Maine south to North Carolina, Georgia, and Alabama. The range extends west as far as Iowa and Minnesota. The eastern milk snake (or intergrade populations) can be expected to occur throughout Virginia.

Range: Scarlet Kingsnake: This snake is found from New Jersey south to southern Florida and the Gulf Coast. It ranges westward to Louisiana, western Tennessee, and western Kentucky. In Virginia, this subspecies has been found primarily in the southeastern portion of the state.

The range of these two subspecies is rather unique and has been concisely explained by Conant (1975). He stated: "The relationship of the Scarlet Kingsnake to the Eastern Milk Snake is highly unusual. In the Cumberland and Interior Low Plateaus of Tennessee and Kentucky, and around the edges of the southern Appalachians, the Scarlet Kingsnake and the Eastern Milk Snake occur together and maintain their identities. Conversely, they intergrade to produce a wide variety of color and pattern combinations in the lowlands and Piedmont from southern New Jersey to northeastern North Carolina. The most frequently occurring variation was formerly recognized as the 'coastal plain milk snake.'"

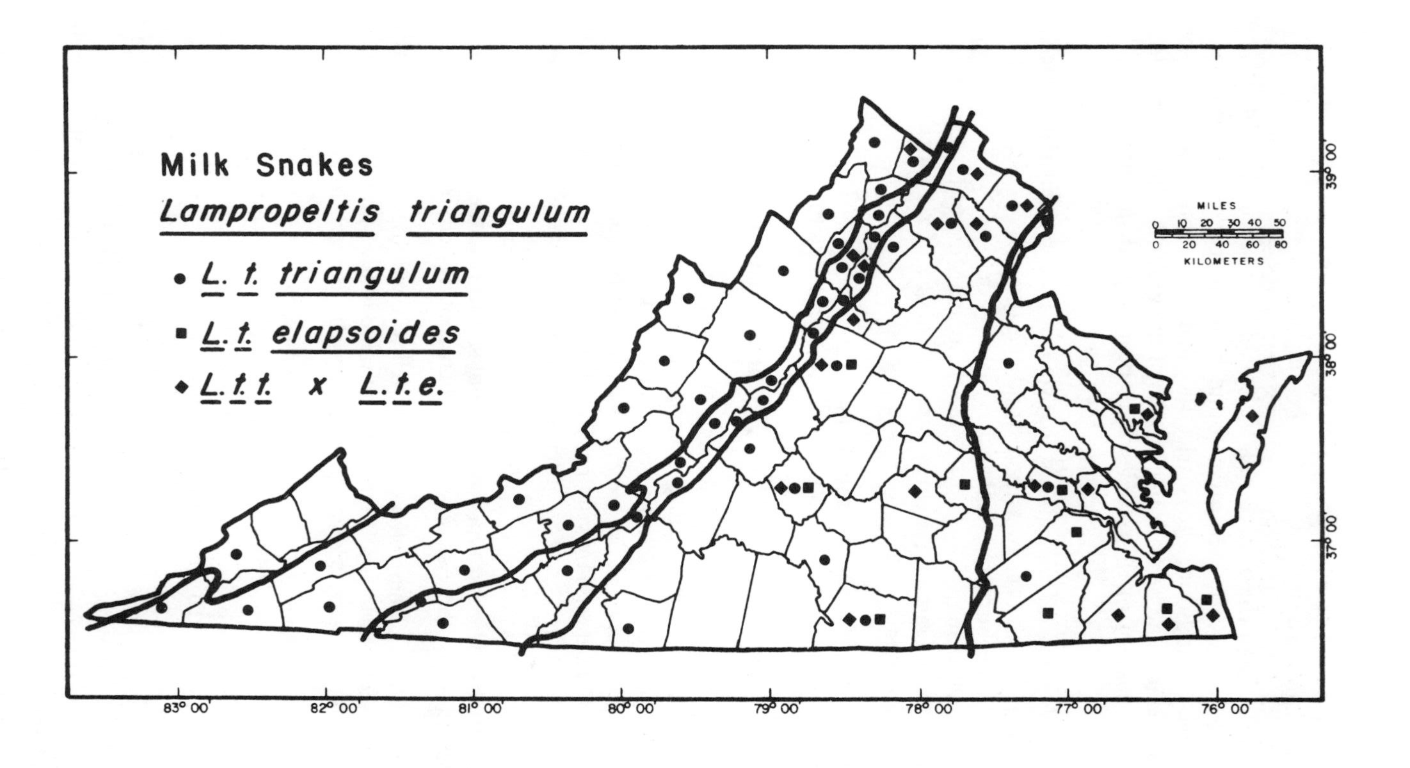
Milk Snakes
Lampropeltis triangulum
L. t. triangulum
L. t. elapsoides
L.t.t. x L.t.e.
MILES
0 10 20 30 40 50
0 20 40 60 80
KILOMETERS
39° 00'
38° 00'
37° 00'
83° 00'
82° 00'
81° 00'
80° 00'
79° 00'
78° 00'
77° 00'
76° 00'

An excellent discussion of the situation in Virginia is presented by Williams (1978).

Habits. Both the eastern milk snake and the scarlet kingsnake are secretive animals and are often found beneath rocks, logs, stumps, and boards. In addition, the scarlet kingsnake is a burrower and spends much of its time beneath the surface of the ground. The systematics and natural history of this species have been treated in detail by Williams (1978).

Reproduction. These snakes are oviparous. Females usually deposit between 4 and 12 eggs in rotting wood and beneath rocks and logs. Some eggs may be buried several inches deep in the soil. The creamy-white, slender eggs are usually laid in early summer. Most of the eggs adhere to one another. Hatching occurs after a period of 2 to 2½ months with newborn individuals ranging from 4 to 8 inches in length.

Longevity. Maximum known age:

Eastern Milk Snake—21 years, 4 months, 14 days (Bowler, 1977).

Scarlet Kingsnake—13 years, 17 days (Bowler, 1977).

Food. Small snakes, lizards, and mice comprise the main food items of these snakes. Earthworms, insects, and small frogs may also be eaten. Uhler, Cottam, and Clarke (1939) analyzed the stomach contents of 19 milk snakes taken in the George Washington National Forest in Virginia. Major food items, by volume, included mice 42%, snakes 26%, songbirds and their eggs 16%, insects 11%, and shrews 5%.

Enemies. The secretive habits of these snakes protect them from many potential predators. Opossums, skunks, and raccoons undoubtedly discover some of these snakes during their rooting activities.

In Captivity. These snakes adapt well to a woodland terrarium as long as sufficient cover is available and suitable food is provided. Their secretive habits, however, keep them out of sight most of the time.

Folklore: The main tale surrounding milk snakes is their supposed habit of milking cows. Milk snakes are frequently observed in and around barns because of the abundance of mice in these areas. They are, however, incapable of performing the action of sucking. Even if they could secure milk in this way, the amount of milk that an individual snake could hold would be so little that a farmer or dairyman would never notice that any was missing. Furthermore, a cow would be highly unlikely to put up with the effects of a snake's needle-sharp teeth during the milking process.

Location of Specimens. AMNH, ANSP, CM, DS, DU, FMNH, FSU, GMU, GMW, MCZ, MLBS, NLU, RWB, UAMNH, UMMZ, USFWS, USNM, VCU.

Mole Kingsnake
(*Lampropeltis calligaster*)

Plate 45

Other Common Names: brown kingsnake, brown snake, blotched kingsnake, mole catcher.

Description. Adult: Yellowish or olive-brown with small reddish-brown blotches down the back, alternating with smaller blotches on the sides. Each blotch has a narrow black border. Belly is yellowish-brown with indistinct brown spots.
Juvenile: Similar to adult but with blotches being dark-edged and more vivid on young snakes and becoming much more indistinct with age.
Scalation: Dorsal scales smooth; 19–23 scale rows; anal plate undivided. Loreal and preocular scales present (fig. 32).

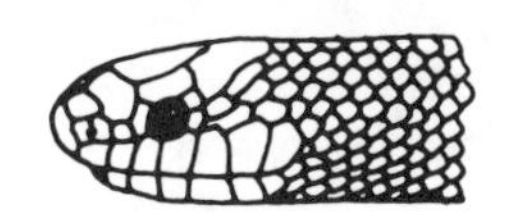

Fig. 32. Head of mole kingsnake, natural size

Size: From about 8 inches at hatching to nearly 4 feet. Most adults are between 2½ and 3½ feet long.
Variation: *Lampropeltis calligaster rhombomaculata* (Holbrook), the mole kingsnake, is the only subspecies occurring in the state.
Similar Species: Sometimes mistaken for corn snake, eastern milk snake, or copperhead. See Table D above.

Habitat. This species is well adapted to farmlands with their mixture of cultivated fields, woodlots, pasture, and weed patches. Mole kingsnakes seem to prefer well-drained land and areas of light soils rather than clays. They are common in some rural villages and even suburbs, but are rarely seen except at night.

Range. The mole kingsnake ranges from southern Maryland to northern Florida and west to Louisiana, Mississippi, and central Tennessee. The range encompasses approximately the eastern two-thirds of Virginia.

Habits. Mole kingsnakes are subterranean and nocturnal and are most often seen when turned up by a plow or when crossing roads at night. Musick (1972), however, noted that they are often seen on the surface of the ground during the day in coastal Virginia. They prowl mole tunnels in search of mice as well as moles and use these tunnels as retreats. They also burrow on their own. The mole kingsnake is generally docile, but its initial defensive reactions are similar to the common kingsnake's. The young ones are very active on defense and will coil and strike so vigorously that they nearly jump off the ground.

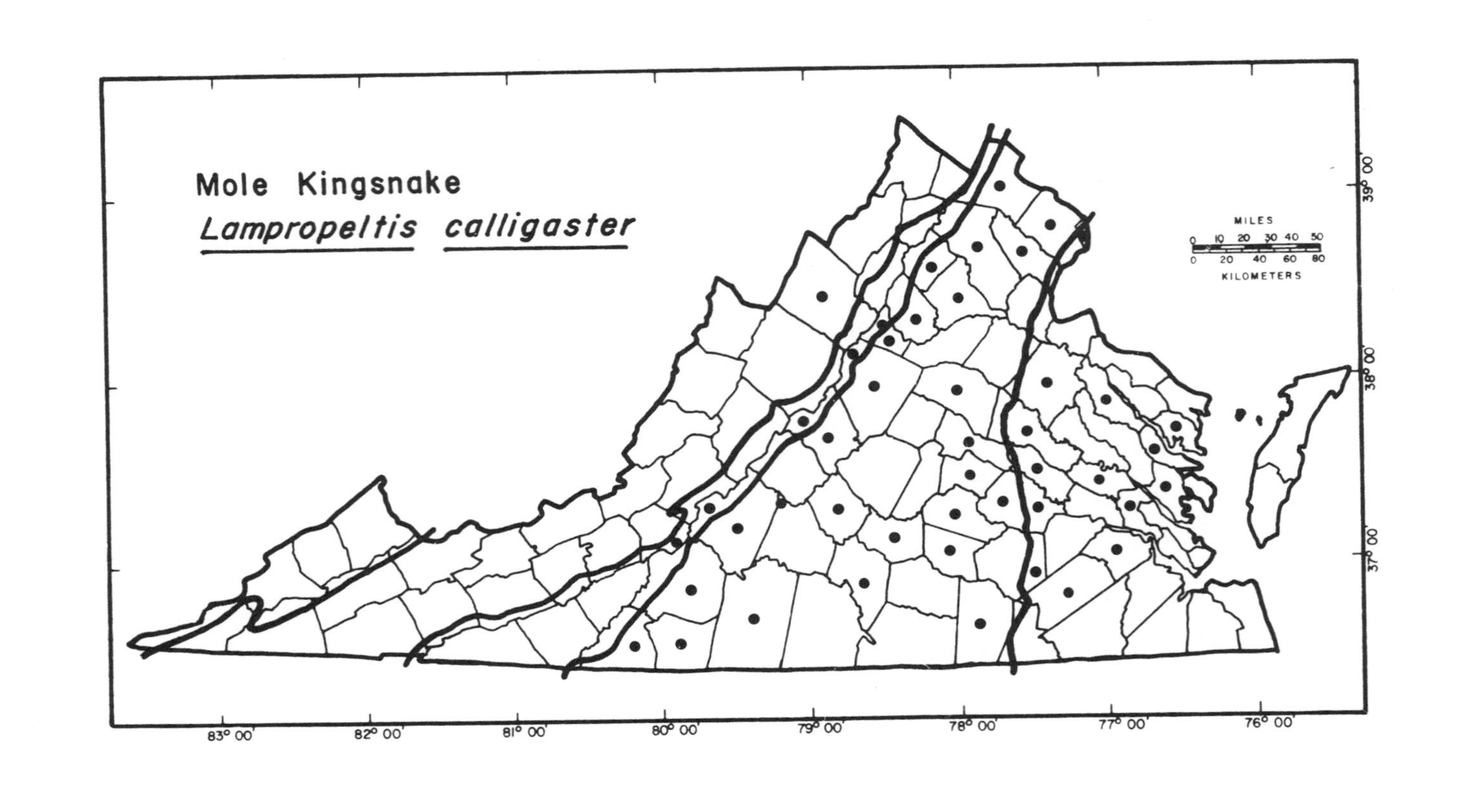

Mole Kingsnake
Lampropeltis calligaster
MILES
0 10 20 30 40 50
0 20 40 60 80
KILOMETERS
39° 00'
38° 00'
37° 00'
83° 00'
82° 00'
81° 00'
80° 00'
79° 00'
78° 00'
77° 00'
76° 00'

Reproduction. Female mole kingsnakes are oviparous and deposit approximately 10 to 16 eggs (Ernst et al., 1985). The eggs tend to adhere to each other in clusters. Hatching occurs in late summer with newborn snakes being approximately 8 inches long.

Longevity. Maximum known age: 6 years, 4 months (Clifford, unpublished).

Food. Mice, snakes, and lizards are favorite foods of adults. Moles and shrews may also be taken on occasion. Young mole kingsnakes have a greater tendency than adults to eat reptiles and will also feed on small frogs and toads. Mole kingsnakes, like other kingsnakes, are constrictors.

Enemies. The beneficial effects of mole kingsnakes are seldom recognized by man. In fact, these snakes are often mistaken for copperheads and are subsequently killed. Quite a few are run over on roads at night and these DOR (dead on road) specimens are sometimes the only indication that the species is present in an area. Natural enemies include raccoons, opossums, and skunks.

In Captivity. Some individuals of this species make excellent captives—they become docile and feed readily. Others under the same conditions are nervous and refuse to eat. The kind of food taken also varies greatly with the individual. Some eat only mammals; others prefer reptiles.

Location of Specimens. AMNH, ANSP, CM, CU, CWM, FMNH, GMU, GMW, LC, MCZ, NVCC, UMMZ, USNM, VCU, VIMS, VPI and SU.

SCARLET SNAKES (Genus *Cemophora*)

Scarlet snakes are related to the kingsnakes (genus *Lampropeltis*) and resemble several of them in coloration and habits. These secretive, semifossorial snakes are small to medium-sized forms that have smooth dorsal scales and an undivided anal plate.

Northern Scarlet Snake (*Cemophora coccinea*)

Other Common Names: false coral snake, candy cane snake. *Plate 46*

Description. Adult: Red and yellow (or white) bands on the back are separated by narrower black bands. Bands are present only on the dorsal surface and do not extend across the belly as in the scarlet kingsnake. Dark dots may occur in the yellow and red areas on older specimens. The belly is whitish or yellowish. The snout is red, pointed, and projects beyond the lower jaw. Head is barely distinct from the neck.

Juvenile: Similar to adult.
Scalation: Dorsal scales smooth; 19 scale rows; anal plate undivided. Loreal and preocular scales present (fig. 33).

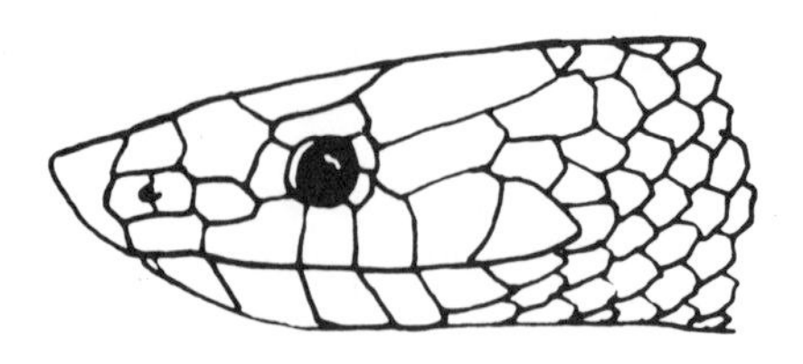

Fig. 33. Head of northern scarlet snake, 2× natural size

Size: From about 5 inches at hatching to slightly over 30 inches. Usual adult size is about 12 to 18 inches.
Variation: *Cemophora coccinea copei* Jan, the northern scarlet snake, is the only subspecies occurring in the state.
Similar Species: The scarlet kingsnake and eastern milk snake have strong and distinct belly markings. The coral snake, a poisonous species not found in Virginia, also has similar colors but in a different sequence.

Habitat. Northern scarlet snakes usually live in areas of light soils, particularly the "sandy lands." The habitat is often covered with pine or scrub oaks, but may be more open. They are seldom seen except when uncovered by plowing, digging, or by moving logs, boards, bark, and rocks which are used as hiding places. They sometimes cause considerable commotion in residential areas when discovered beneath a flowerpot or unearthed in a garden.

Range. The scarlet snake is primarily a southeastern species and ranges from New Jersey to northern Florida and west into Texas and Oklahoma. The main range extends northward into southern Illinois and southern Indiana. Several disjunct populations exist outside of the main range. This species is found in the eastern half of Virginia with an apparent disjunct population on the Virginia-West Virginia border. Further studies may show that this species penetrates the highlands via river valleys such as the James River (Hoffman, 1977).

Habits. These snakes are accomplished burrowers that usually prowl above ground only at night. Heavy rains may force them to the surface where they seek cover under a variety of objects. These same objects often function as cover for the scarlet snake's prey and thus also serve as hunting areas.

Reproduction. The oviparous females deposit from three to eight elongate, whitish eggs in moist, decaying logs or in moist soil. The eggs are between 1¼ and 1½ inches long and are normally laid during June and

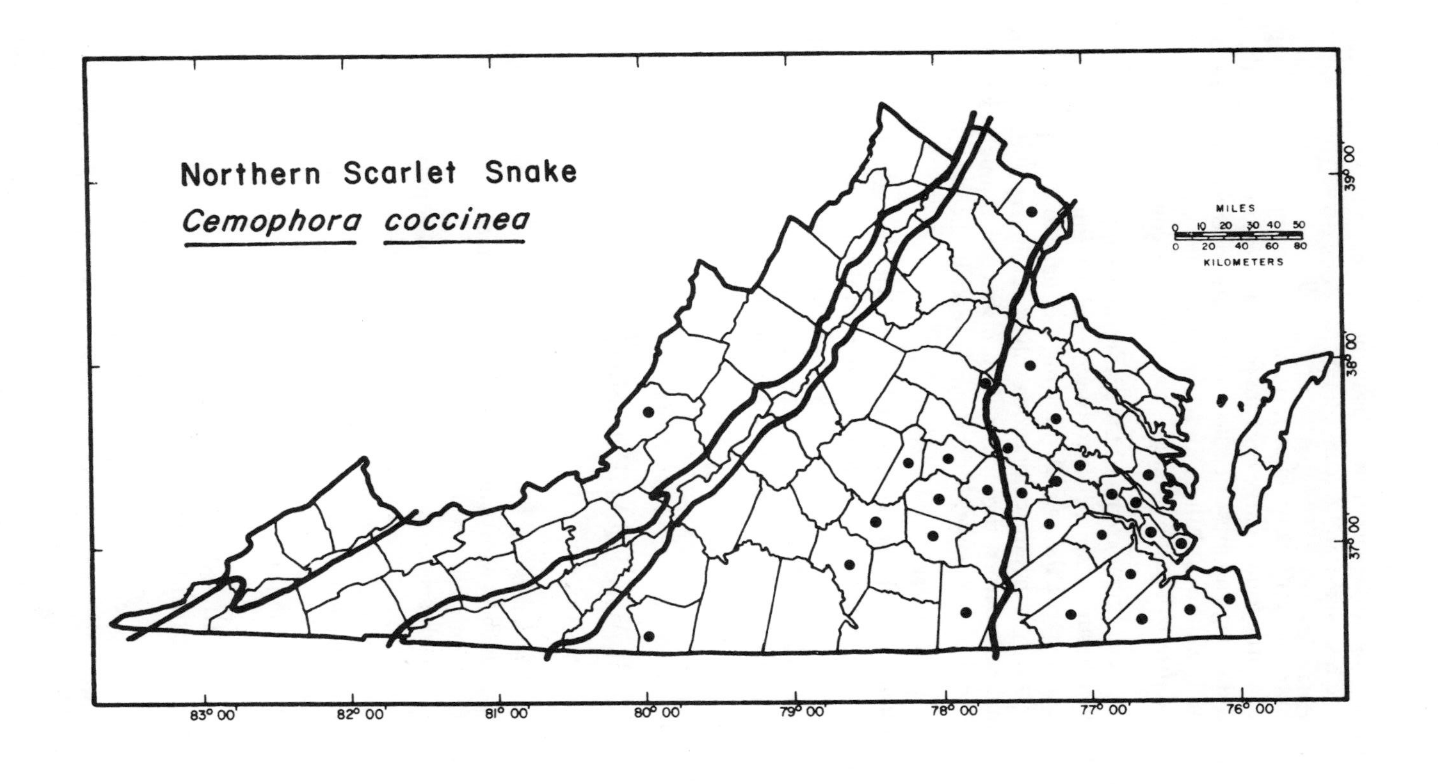
Northern Scarlet Snake
Cemophora coccinea
MILES
0 10 20 30 40 50
0 20 40 60 80
KILOMETERS
39° 00'
38° 00'
37° 00'
83° 00'
82° 00'
81° 00'
80° 00'
79° 00'
78° 00'
77° 00'
76° 00'

July. Hatching takes place from midsummer to early fall. Newborn scarlet snakes are between 5 and 6 inches long.

Food. Lizards (especially skinks), small mice, young snakes, and reptile eggs are favorite foods. Large eggs are pierced by enlarged rear teeth, and through chewing and pressure, the contents are squeezed into the snake's throat and swallowed (Palmer and Tregembo, 1970). Other prey include frogs and insects. Most prey animals are killed by constriction.

Enemies. The bright colors of the scarlet snake apparently evolved as a defensive measure against predators with color vision. The colors could momentarily startle the animal, allowing the snake to attempt an escape. Another theory is that the coloration evolved as mimicry of the dangerous coral snake. Predators, through evolved instinct or through experience, would avoid such a snake. Most humans react in the desired manner in both theories.

In Captivity. Scarlet snakes do well in captivity if acceptable foods are available. Some individuals will eat only reptile eggs; others only skinks. Thus, these individuals generally present more of a problem than the specimens that will feed on mice. A screen-covered terrarium containing an inch or two of wood chips makes a suitable home.

Location of Specimens. CM, CWM, GMW, MCZ, UF, UMMZ, USNM, VCU, VIMS, VMNH.

BLACK-HEADED SNAKES
(Genus *Tantilla*)

The snakes of this genus are all small, with none reaching a length of 24 inches. Most adults are less than 12 inches long. The dorsal scales are smooth, and the anal plate is divided.

Black-headed snakes possess a mild venom and short, extremely small, rigid fangs located at the rear of the upper jaw. This primitive apparatus is used in subduing prey. Unlike the sophisticated pit vipers that inject their venom, the rear-fanged snakes usually must grasp and chew to ensure that their venom enters the prey. Black-headed snakes can be considered completely harmless to humans.

Southeastern Crowned Snake
(*Tantilla coronata*)

Other Common Names: black-headed snake, ground snake. *Plate 47*
Adult: Dorsal surface is light brown to reddish-brown with no markings. Head is black. Light band or collar present across back of head bordered by a black band 3 to 5 scales wide. Belly white, often with a pinkish or yellowish cast. Flat, rather blunt head; small eyes; slender body.

Juvenile: Similar to adult.
Scalation: Dorsal scales smooth; 15 scale rows; anal plate divided. Preocular scale present; loreal scale absent (fig. 34).

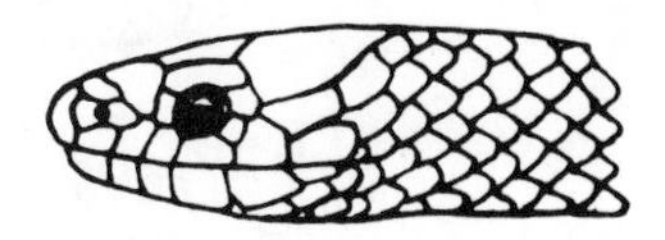

Fig. 34. Head of southeastern crowned snake, 3× natural size

Size: From about 3 inches at hatching to slightly over 12 inches. Adults are usually between 8 and 10 inches.
Variation: No subspecific variation is recognized.
Similar Species: See Table E above. No other snakes have both white and black collars.

Habitat. Crowned snakes are most often found on dry, wooded, and rocky hillsides. They are occasionally found in backyards. An abundance of stones and logs are required to provide cover for these little snakes. This species has also been found in moist woodlands and near swamps and streams.

Range. The southeastern crowned snake is another species whose range is restricted to the southeastern United States. The range extends from south-central Virginia and southern Illinois to South Carolina, then southwestward to the Gulf Coast. It extends westward to Louisiana, western Tennessee, and western Kentucky. In Virginia, crowned snakes have been found along the eastern edge of the Blue Ridge and in the extreme southwestern portion of the state. This species has been officially classified as "Status Undetermined" in Virginia (Mitchell, 1991).

Habits. This is a very secretive species, seldom being found in the open. Logs, stumps, loose bark, stones, rocks, and crevices are used as hiding places. They may also burrow into the soil. The somewhat wedge-shaped head is a useful tool in the snake's maneuverings through these environments. The species is probably more widespread and common in the southern Piedmont, but it is so secretive that it is seldom seen.

Reproduction. Among the family Colubridae, the genus *Tantilla* is unique in that some species, including *Tantilla coronata,* have a vestigial left oviduct yet have functional left and right ovaries (Clark, 1970; Aldridge, 1992). The left oviduct terminates 2–3 mm anterior to the vagina. Mating occurs in the late summer and fall and/or the following spring. Sperm from late summer and fall matings overwinter in the posterior oviduct. Female crowned snakes lay between one and six eggs in early sum-

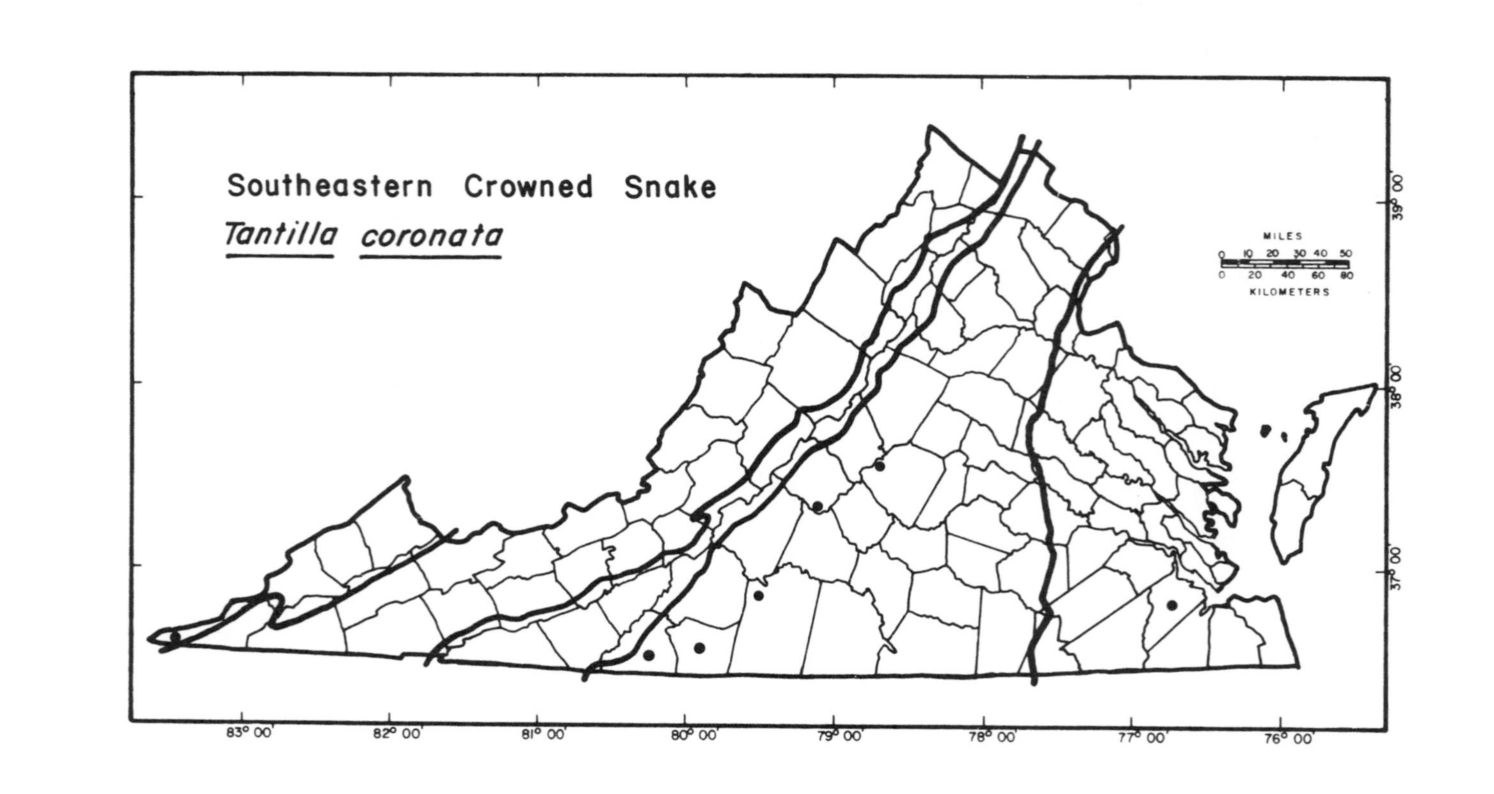
Southeastern Crowned Snake
Tantilla coronata
MILES
0 10 20 30 40 50
0 20 40 60 80
KILOMETERS
39° 00'
38° 00'
37° 00'
83° 00'
82° 00'
81° 00'
80° 00'
79° 00'
78° 00'
77° 00'
76° 00'

mer. The eggs are placed in rotting logs, stumps, or other decaying material. The young snakes are about 3 inches long at hatching.

Food. Small invertebrates such as centipedes and insect larvae are the primary prey. They may be partially immobilized by the snake's venom while being swallowed.

Enemies. While the crowned snake's mild venom and primitive injection apparatus probably has little defensive value, the snakes maintain their populations despite a small reproductive potential. Their secretive habits undoubtedly protect them from many potential predators.

In Captivity. These are not the most desirable of pets. Since they stay hidden most of the time, observation is limited. They can be kept in a dry woodland terrarium.

Location of Specimens. CM, RMWC, USNM.

PIT VIPERS

Family Viperidae
Subfamily Crotalinae

Two groups of poisonous snakes occur in the United States—the pit vipers and the coral snakes. The range of the coral snakes extends from North Carolina south to southern Florida and west to Texas and Mexico; thus their range does not extend into Virginia, even though Barringer (1892) noted that the range of this snake "is from Southern Virginia around the South Atlantic States to the Gulf; up the Mississippi Valley to Missouri, and thence to Texas." One or more kinds of pit vipers, however, are found in every state except Maine. Five varieties—northern copperhead, southern copperhead, eastern cottonmouth, timber rattlesnake, and canebrake rattlesnake—occur in Virginia. A sixth—the pygmy rattlesnake (*Sistrurus miliarius*)—has been taken in Virginia on only one occasion in the vicinity of Northwest River, Chesapeake City County, in November 1957 (Tobey, 1960; Witt, 1962; Palmer, 1971). The nearest population is south of Albemarle Sound, North Carolina. Since no additional individuals have been reported, it may be assumed that this single specimen either escaped or was released near the site of capture, although Hardy (1972) referred to a disjunct population in southeastern Virginia. Brady (1927) reported the discovery of a snake from the Dismal Swamp whose description fit that of the eastern diamondback rattlesnake (*Crotalus adamanteus*). However, no additional specimens have been reported.

Pit vipers get their name because they possess a pair of heat-sensing pits between their eyes and nostrils. In addition, they possess vertical pupils and a more or less triangular-shaped head that is very flat on top and meets the side of the face at a sharp angle. All of the nonpoisonous snakes in Virginia have round pupils, and their head and facial region is much more rounded. Although most of our pit vipers have a triangular-shaped head, this character alone should not be used to identify these snakes.

The heat-sensing pits are valuable aids particularly when these snakes are feeding upon warm-blooded prey such as mammals or birds that continuously give off a certain amount of body heat. Each pit consists of two cavities, an outer and an inner, separated by a membrane. The pits are able to detect temperature differences of as little as 1°C higher or lower than that of the background. The range is limited to about

12 inches in adult rattlesnakes; objects of different temperatures can be localized within this zone of sensitivity. The pits thus enable the snakes to strike very accurately at the source of heat. Since both snakes and prey are normally most active at night when the ambient or surrounding air temperature is cooler, this sensory apparatus is even more efficient.

Pit vipers strike by swiftly straightening the front portion of their body from its S-shaped position. The snake recovers its original position in a moment and is ready to strike again. Contrary to popular belief, it is not essential for these snakes to be in a coiled position in order to strike. Snakes can strike for a distance equal to between 1/3 and 1/2 of their total length.

All pit vipers have a pair of hollow fangs near the front of their mouth. These fangs represent modified teeth. They are hinged at their base so that when the mouth is closed, they can be folded back along the roof of the mouth. A protective tissue sheath encloses each fang when the mouth is closed. This sheath is retracted when the mouth is opened. Whenever a fang becomes damaged or broken, a new one will grow in its place. This type of continual replacement is true not only for the fangs but for every tooth in the snake's mouth.

The venom, a yellow viscid liquid which looks somewhat like orange juice and dries into pale yellow crystals, is produced in glands near the point where the upper and lower jaws join. As the snake strikes and embeds its fangs in the prey, the muscles surrounding the poison sacs contract and squeeze the venom along ducts leading to the base of the fangs. The venom then travels through the hollow fangs and out a small opening at their tip into the prey. The entire mechanism is very similar to a hypodermic syringe, with the venom actually being injected into the prey. This entire process takes place very rapidly as the snake strikes its prey and draws back quickly.

The venom of pit vipers is a hemotoxic (hemolytic) venom which serves to break down and destroy blood cells and other tissues and lowers the ability of the blood to coagulate or clot, thus resulting in hemorrhage throughout any portion of the circulatory system penetrated by the poison. This is in contrast to the neurotoxic venom of coral snakes which attacks the central nervous system of the victim.

The severity of a snakebite is dependent upon the type, size, and condition of the snake as well as the place of the bite on the victim and the size and physical condition of the victim. Furthermore, a snake which had recently expended much of its venom in securing food cannot do as much damage as a similar-sized snake with a full supply of venom. Most individuals are struck below the elbow or below the knee. Fortunately, these distal portions of the arms and legs are not as richly supplied with blood as are most other portions of the human body. Therefore, it is often possible to extract much of the venom before it is transported via the blood and lymph channels to the vital organs.

Various theories abound as to the correct treatment for snakebite. In most cases, a constriction band should be immediately applied above the wound. The victim should not be given any alcoholic beverages and strenuous exercise should be avoided, if possible. Alcohol and exertion only help spread the poison to the rest of the body. If the victim can be transported to a doctor or hospital within approximately 30 minutes, no other field treatment is necessary. Medical personnel will inject antivenin to counteract the effects of the venom. If medical assistance is not available within 30 minutes, the area of the bite should be cooled and as much venom as possible should be removed by suction. The cooling can be accomplished either by submerging the affected part in a cold mountain stream or spring or by carefully packing and cooling it in ice. By lowering the temperature, the circulation of the blood is slowed, thus slowing the spread of the venom in the blood. If ice is used, care must be taken to prevent the affected area from freezing and/or becoming frostbitten. Instances have been recorded where overzealous rescuers have caused more damage to the victim than was caused by the original snakebite. Small incisions should be made through each fang mark and suction should be applied, either by mouth or with the suction device contained in a snakebite kit. The incisions, which should not be very deep, must be made with care so as not to cut so deep that major blood vessels, nerves, and tendons are damaged. As long as there are no open sores in the mouth, the venom and tissue fluids may be extracted by using the mouth for suction.

Death from snakebite in the United States is rare. In an average year, more persons are killed by lightning than by snakebite. Records indicate approximately 5,000 snakebites per year in the United States with 10–15 fatalities each year. Many of these bites result from people handling the snakes. Rattlesnakes account for most snakebites in the United States and for 80 to 90 percent of the fatalities.

The first quantitative study of snakebite in the United States was undertaken by Willson (1908). Out of 566 cases, there were 408 cases of rattlesnake bite, 97 cases of copperhead bite, 53 cases of water moccasin bite, and 8 cases of coral snake bite. Hutchinson (1929; 1930) reported that approximately 4 percent of the 607 cases occurring in the United States during 1928 had taken place in Virginia. Wood (1954) noted the following species of snakes involved in 134 cases of snakebite occurring in Virginia primarily between 1941 and 1953: copperhead, 119; cottonmouth, 1; timber rattlesnake, 12; canebrake rattlesnake, 2. Data recorded by Wood showed that 96 percent of the cases occurred in a five-month period from May through September with a peak incidence of 32 percent in July. In Virginia, 52 percent of the persons bitten by poisonous snakes were under 16 years of age. Most children between 2 and 8 years of age were bitten near their homes or on farms near their homes, while the majority of children between 9 and 15 years of age were bitten in areas remote from their homes such as in woods and fields. Parrish (1957) compiled records of 71 instances of snakebite mortality in the United States during the period 1950–54. Parrish (1963; 1966) reported 15 deaths in the United States during 1958 and 14 deaths during 1959. Of these deaths, about 6.6 percent were attributable to cottonmouths, 77.0 percent to rattlesnakes, and 1.6 percent to coral snakes; 14.8 percent were unidentified. Almost half of the fatalities are in persons less than 20 years of age, the high mortality rate being partially due to the greater ratio of venom to body weight. No deaths were recorded in Virginia during the period 1950–59. During 1959, the incidence of snakebite in Virginia was 5.47 per 100,000 persons. For comparison, the incidence was 18.79 for North Carolina, 11.29 for West Virginia, 4.71 for Kentucky, 2.21 for Tennessee, and 1.35 for Maryland.

COPPERHEADS AND COTTONMOUTHS (Genus *Agkistrodon*)

This genus contains medium- to large-sized snakes that have weakly keeled scales and an undivided anal plate. Large individuals are generally heavy-bodied and robust. Gloyd and Conant (1989) presented a monographic review of this genus.

Copperhead (*Agkistrodon contortrix*)

Plates 48–49

Other Common Names: highland moccasin, pilot snake, chunk head, upland moccasin.

Description. Adult: Series of reddish-brown hourglass crossbands on a ground color ranging from pale reddish or pinkish-brown to tan. Crossbands may be strongly constricted in middle of back, appearing from the side as a dark triangle with its apex directed upwards. Head usually a coppery color. Belly ranges from pinkish to tan with irregular dark spots and blotches.
Juvenile: Paler than adult; narrow dark line through the eye; yellow tail tip.
Scalation: Dorsal scales weakly keeled; 23 scale rows; undivided anal plate. Loreal and preocular scales present (fig. 35). Single row of scales under most of tail.

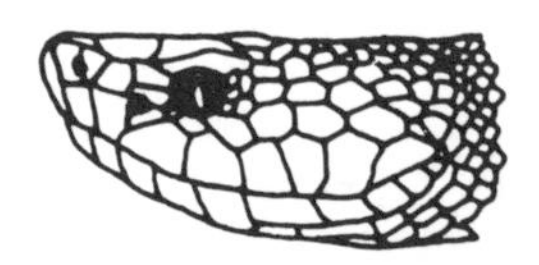

Fig. 35. Head of northern copperhead, natural size

Size: From 8 to 10 inches at birth to more than 4 feet. Most adults are generally between 2 and 3 feet.
Variation: *Agkistrodon contortrix mokasen* Beauvois, the northern copperhead, is the only subspecies occurring in the state.
Similar Species: See Table D above. Young are often confused with the young of the cottonmouth, both of which have yellow or yellowish-green tail tips and a coppery head.

Habitat. The copperhead, which is commonly known as the "highland moccasin," is most often found in mountainous rocky and hilly country. Rocky ledges, rock outcrops, and stone walls are favorite sites, although individuals have also been observed in cultivated fields, meadows, canebrakes, streams, and swamps. Musick (1972) noted their association with wild blueberry shrubs. Abandoned sawdust piles and slab heaps also provide a suitable habitat. Hardy (1972) noted that this species occurs in the immediate vicinity of barrier beach ponds and coastal marshes and that a single specimen was captured on a sandy beach in a wet intertidal zone in the Chesapeake Bay region.

Range. The copperhead is found throughout most of the eastern United States from Massachusetts, New York, Illinois, and Iowa south to northern Florida and the Gulf Coast and west into Kansas, Oklahoma, and Texas. This species occurs throughout Virginia. Some snakes with traces of characteristics of the southern copperhead (*Agkistrodon contortrix contortrix*) were reported from the Dismal Swamp by Burger in 1958–59.

Habits. Copperheads are normally not aggressive snakes. A number of persons, including one of us (Linzey), have come within inches of a well-camouflaged copperhead without causing it to be disturbed.

The yellow or yellowish-green tail tips of juvenile copperheads (and cottonmouths) are thought by some to be used as lures, particularly for small frogs. Evidence in support of this theory has been presented by Ditmars (1907) and Neill (1948*b;* 1960).

In a study of thermal ecology, Sanders and Jacob (1981) noted that copperheads were diurnal in the spring, nocturnal in the summer, and diurnal again in the fall. In mountainous areas copperheads gather in the vicinity of overwintering "dens" with blacksnakes and timber rattlesnakes during the fall months. Most dens are located on wooded mountainsides with southern exposures. Dens consist of weathered outcroppings of rock with deep crevices that allow the snakes to crawl below the frost line. Finneran (1953) reported that in some regions females gather at dens in late summer and early fall, presumably near the time of completion of the gestation period. Copperheads living in the Coastal Plain are reported to overwinter singly (Neill, 1948). In Virginia, copperheads have been observed as early as April 16 and as late as December 12 (Wood, 1954). The life history and ecology of the copperhead throughout its range were treated in detail by Fitch (1960).

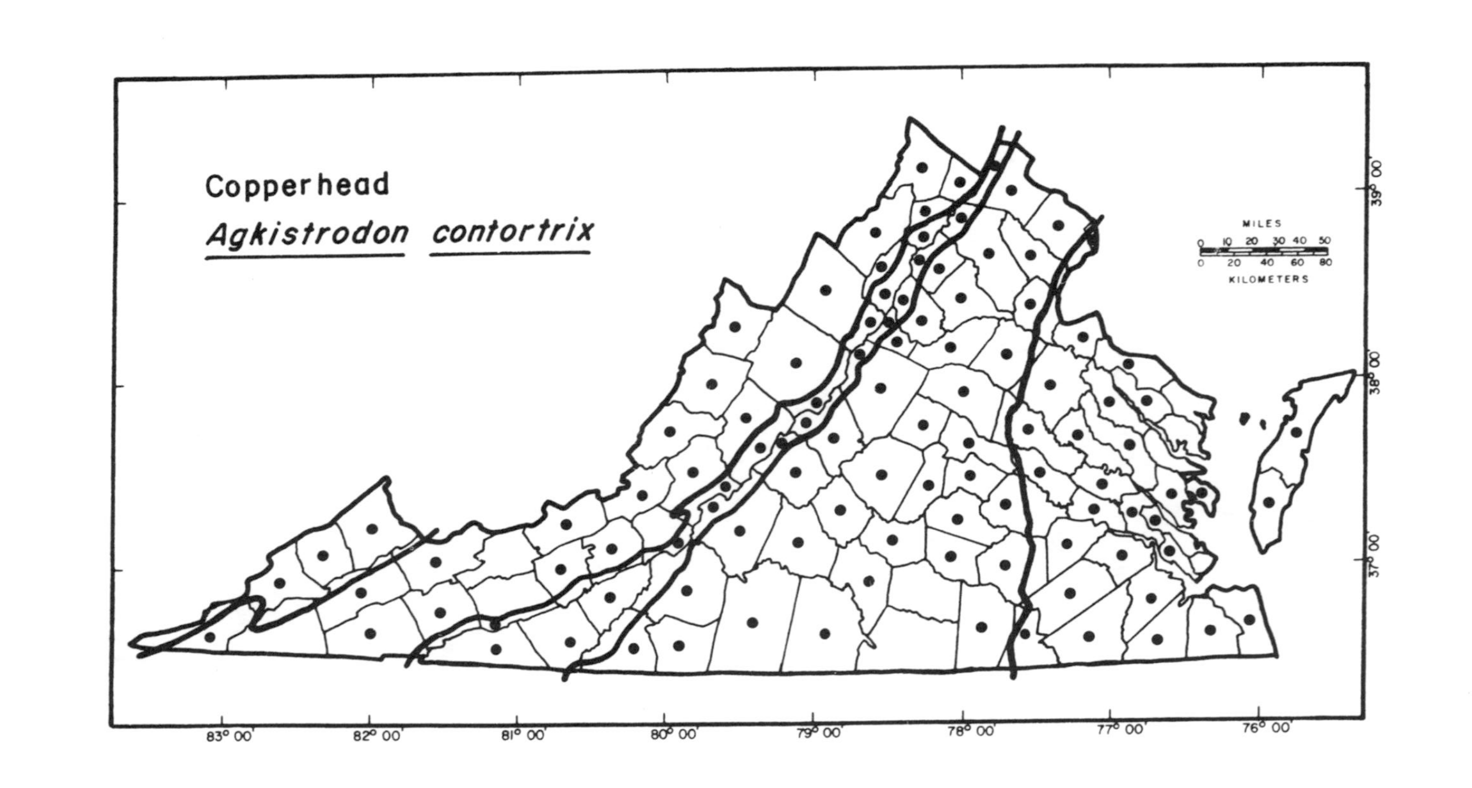
Copperhead
Agkistrodon contortrix
MILES
0 10 20 30 40 50
0 20 40 60 80
KILOMETERS
39° 00'
38° 00'
37° 00'
83° 00'
82° 00'
81° 00'
80° 00'
79° 00'
78° 00'
77° 00'
76° 00'

Reproduction. Copperheads are ovoviviparous snakes with females giving birth to between 2 and 17 young during August and September. Hoffman (1945) recorded a female from Alleghany County that gave birth to 4 young on August 20. Young copperheads are usually between 8 and 10 inches long at birth. In a study of the copperhead in Kansas, Fitch (1960) recorded a biennial reproductive cycle. Courtship and mating were discussed by Schuett and Gillingham (1988).

Longevity. Maximum known age: Northern Copperhead—29 years, 10 months, 6 days (Bowler, 1977).

Food. Copperheads feed mainly on small mammals such as mice, shrews, and young rats. Other prey may include small birds, frogs, toads, salamanders, and invertebrates. Saylor (1938) recorded hairy-tailed moles from Augusta County. Uhler, Cottam, and Clarke (1939) examined 72 copperhead stomachs from the George Washington National Forest in Virginia. Mice comprised 31% of the total volume of food consumed, caterpillars and moths 21%, shrews 17%, salamanders 10%, moles 5%, and birds 4%. The stomach of a copperhead from the Dismal Swamp contained a short-tailed shrew (Rageot, 1957). In some areas, young copperheads feed extensively on ringneck snakes (Fitch, 1960).

Enemies. These snakes have relatively few enemies. Kingsnakes will feed upon them if the occasion arises. Man is undoubtedly the greatest enemy, with most persons killing these snakes whenever and wherever they are found. Unfortunately, a number of nonpoisonous snakes are killed each year by individuals who mistake them for a copperhead.

In Captivity. Due to their venomous nature, it is not recommended that the amateur snake collector attempt to maintain these snakes in captivity.

Location of Specimens. AMNH, ANSP, ART, CM, CU, CWM, DS, DU, FMNH, GMU, GMW, HSH, JHU, LC, MCZ, MLBS, MSWB, NVCC, ODU, RWB, SDSNH, SNP, TAMU, UK, UM, UMMZ, USFWS, USNM, VCU, VIMS, VMNH, VPI and SU.

Cottonmouth
(*Agkistrodon piscivorus*)

Plate 50

Other Common Names: water moccasin, lowland moccasin, stub-tailed moccasin.

Description. Adult: Olive-brown to brown above with dark crossbands on sides and back. Crossbands are usually wider at their base and narrower at center of back. Pattern most evident in juveniles with some large adults becoming almost uniformly slate black. Belly usually lighter with irregular dark stippling and blotches. Single row of scales on underside

of tail, except near tip. May be large and heavy-bodied. Individuals from Virginia Beach (formerly Princess Anne County) "are marked with unusually vivid crossbands and have an abnormally light ground color" (Werler and McCallion, 1951).
Juvenile: Very prominent pattern of brown or reddish-brown crossbands. Coppery-brown head. A broad dark brown band runs through the eye. Tail tip may be yellow or yellowish-green.
Scalation: Dorsal scales weakly keeled; 25 scale rows; anal plate undivided. Preocular scale present; loreal scale absent (fig. 36).

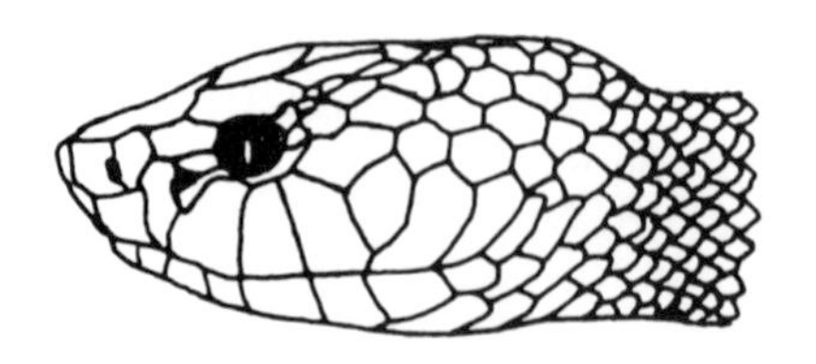

Fig. 36. Head of eastern cottonmouth, natural size

Size: From about 8 inches at birth to more than 6 feet. Most adult cottonmouths are usually between 2½ and 4 feet.
Variation: *Agkistrodon piscivorus piscivorus* (Lacépéde), the eastern cottonmouth, is the only subspecies occurring in the state.
Similar Species: See Tables A and B above. Adults are frequently confused with nonpoisonous water snakes. Young cottonmouths with their coppery-brown head and yellow tail tip are often mistakenly identified as young copperheads.

Habitat. These are semiaquatic snakes which are most often found near water but occasionally may be found in fields and other dry areas. One of us (Linzey) has had the unnerving experience of coming within inches of stepping on a coiled cottonmouth lying on the far side of a log in a dry field quite some distance from any water. Fortunately, it was early on a cool autumn day and the snake was somewhat sluggish.

Cottonmouths prefer such areas as swamps, marshes, and drainage ditches. They are also commonly found along the edges of lakes, ponds, and slow-moving streams and rivers. They may inhabit brackish water areas (Neill, 1958). Most individuals are observed basking on roots or logs in these areas.

Range. The cottonmouth ranges from Virginia to the Florida Keys and west to central Texas and central Oklahoma. Its range extends north to Missouri, southern Illinois, and western Kentucky. This species is found only in the southeastern portion of Virginia. Colonies exist as far north and west as southeastern Chesterfield County. Musick (pers. comm.) noted that an isolated population exists in a swamp within the Newport News City Park. Cottonmouths may also be found on the Virginia shore

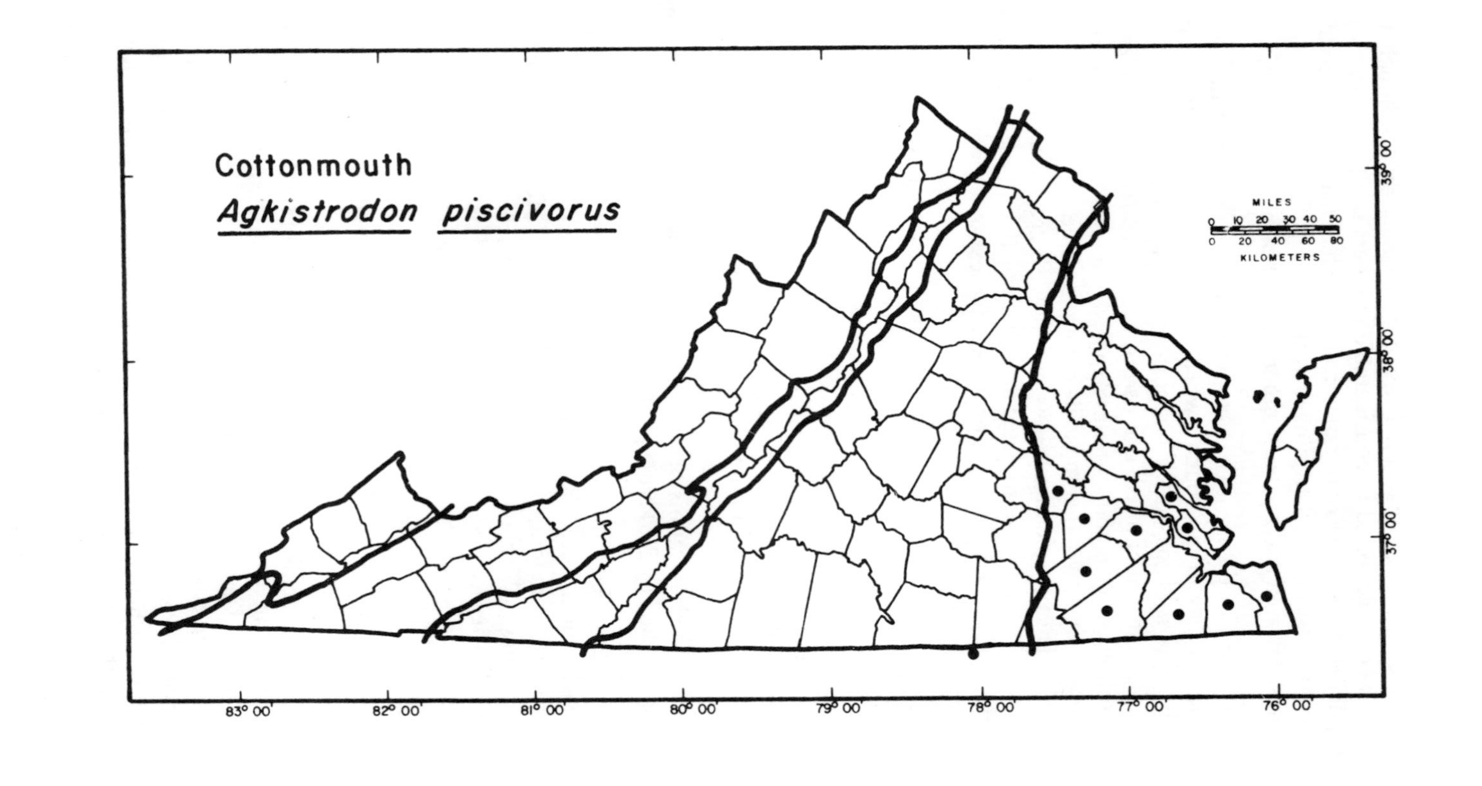
Cottonmouth
Agkistrodon piscivorus
MILES
0 10 20 30 40 50
0 20 40 60 80
KILOMETERS
39° 00'
38° 00'
37° 00'
83° 00'
82° 00'
81° 00'
80° 00'
79° 00'
78° 00'
77° 00'
76° 00'

of Lake Gaston. Hoffman (*in* Russ, 1973) stated that although these snakes were once found in considerable numbers in the area east of the fall line, they are now only rarely found in the Dismal Swamp and in military reserves and other refuges in that region. They are common in the Back Bay area and are fairly numerous in Seashore State Park.

Habits. One of the most characteristic habits of the cottonmouth is its tendency to "stand its ground" when alarmed rather than to escape as quickly as possible as do the nonpoisonous water snakes. During these times, the cottonmouth will coil its body and threaten the intruder with its mouth wide open and its fangs exposed. The open mouth reveals the white lining that gives this snake its common name.

The conspicuous yellow or yellowish-green tail tip of juvenile cottonmouths can be manipulated in such a way that it resembles a worm and may thus aid in luring prey toward the snake (Wharton, 1960; Neill, 1960).

In Virginia, these snakes have been observed from April to December. On one occasion, they have even been found sunning at Back Bay during the first week of January. Instances of cottonmouths apparently migrating from the bayside swamps of barrier beaches to the mainland in late October and early November have been reported (Wood, 1954). As many as 50 snakes have been observed swimming across Back Bay in the vicinity of the wildlife refuge in a period of as little as three or four days.

The natural history of the cottonmouth throughout its range has been studied by Burkett (1966).

Reproduction. Like the copperhead, these snakes undergo ovoviviparous development. Breeding takes place during the spring with birth usually occurring from mid-August to mid-September. Mean litter size for 24 gravid females in Virginia is 7.7 (Blem, 1981). Each of the young snakes is brightly patterned and possesses a yellowish tail tip. Blem (1981, 1982) found 83% of mature females in Virginia were gravid during the breeding season.

Longevity. Maximum known age: 13 years, 1 month, 24 days (Bowler, 1977).

Food. A wide array of prey species are consumed by these semiaquatic snakes. Food items recorded have included fish, frogs, salamanders, lizards, turtles, snakes, birds, and small mammals.

The cottonmouth is fully capable of opening its mouth and biting while beneath the surface of the water. Only in this way could it secure its aquatic prey such as fish. When the snake captures a fish, it may remain in the water while swallowing its catch, or it may carry the fish up onto the bank where it can manipulate its dinner for easier swallowing (Bothner, 1974).

Enemies. Due to their aggressive disposition and venomous nature, these snakes have few enemies. Kingsnakes may occasionally kill some cottonmouths. Great blue herons and largemouth bass have also been known to prey on these snakes. Many are killed by fishermen and others who come upon them. Their general resemblance to nonpoisonous water snakes causes the death of many nonpoisonous individuals by persons who are either unable or unwilling to distinguish between them.

In Captivity. These snakes should not be kept in captivity except by experienced and knowledgeable herpetologists. They retain their aggressive behavior, and great caution must be exercised whenever they are handled.

Location of Specimens. CM, ESU, GMW, MCZ, RFC, SDSNH, UK, USFWS, USNM, VCU, VIMS, VMNH.

RATTLESNAKES
(Genus *Crotalus*)

Rattlesnakes are probably the most familiar kind of poisonous snake to most persons. They are medium- to large-sized pit vipers that are characterized by having a rattle at the end of their tail. All of these snakes possess keeled dorsal scales, an undivided anal plate, and vertical pupils.

Most of the 26 different kinds of rattlesnakes found in the United States inhabit the southwestern portion of the country. Only one species (*Crotalus horridus*) occurs in Virginia. This species consists of a northern and a southern subspecies—the timber rattlesnake and the canebrake rattlesnake, respectively.

Pisani, Collins, and Edwards (1973) concluded that no distinct subspecies of *Crotalus horridus* could justifiably be recognized. Tobey (1979), however, referred to the canebrake rattlesnake as the southeastern population of the timber rattlesnake. He stated: "The recent decision to abolish the subspecies (Pisani et al., 1973) falls hardest on Virginia where the canebrake rattler and timber rattler populations are most distinct and widely separated geographically. We must recognize, however, that this gap may have been produced by man within the past 200 years. For practical purposes, within Virginia, we will continue to use the common name of the former subspecies and leave the technical debate to others." Brown and Ernst (1986) presented evidence suggesting that, on the basis of differences in adult size and pattern, two subspecies occur in Virginia. Pending further studies of this species in Virginia, we have decided to continue to recognize both subspecies.

Timber Rattlesnake
(*Crotalus horridus*)

Plates 51–53

Other Common Names: black rattlesnake, eastern rattlesnake, banded rattlesnake, velvet-tail rattler.

Description. This species consists of two subspecies which differ considerably in external appearance; hence they are treated separately.
Adult:
Timber Rattlesnake: Exists in two color phases.
Yellow phase: Yellowish or yellowish-gray with series of dark brown or black V-shaped crossbands. Tail dark brownish-black. Belly light with irregular black markings.
Black phase: Considerable black or dark brown stippling that partially or completely obscures yellowish ground color. Dark brown or black V-shaped crossbands. Tail dark brownish-black. Belly light with irregular black markings.
Canebrake Rattlesnake: Pale grayish-brown to flesh pink ground color. Dark brown or black V-shaped crossbands. Middorsal reddish-brown stripe. Dark oblique band runs posteriorly from eye to beyond angle of the jaw. Tail dark brownish-black. Belly light with black markings or stippling.
Juvenile: Basically similar to adult but with very distinct crossbands.
Scalation:
Timber Rattlesnake: Dorsal scales keeled; 23 scale rows; anal plate undivided. Loreal and preocular scales present (fig. 37*a*).
Canebrake Rattlesnake: Dorsal scales keeled; 25 scale rows; anal plate undivided. Loreal and preocular scales present (fig. 37*b*).

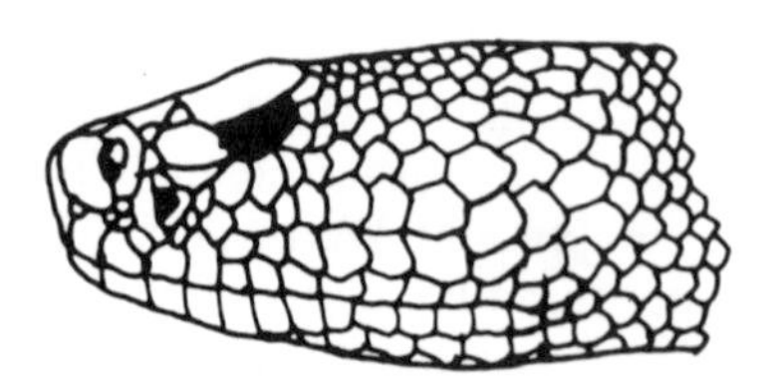

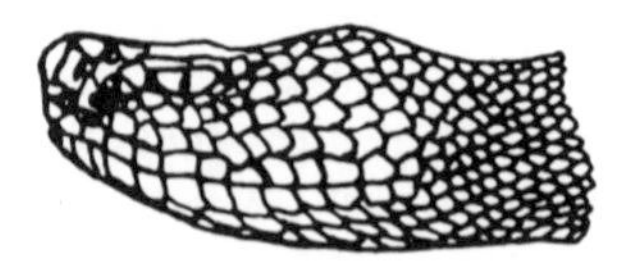

Fig. 37. a, *Head of timber rattlesnake, natural size;* b, *head of canebrake rattlesnake, natural size*

Size: From about 12 inches at birth to over 6 feet. Most adults are usually between 3 and 5 feet long. The largest canebrake rattlesnakes ever reported from Virginia were two 72-inch individuals found in Princess Anne County (Virginia Beach) and near Lake Drummond in the Dismal Swamp (Werler and McCallion, 1951; Martin and Wood, 1955).
Variation: Two subspecies are recognized in Virginia.

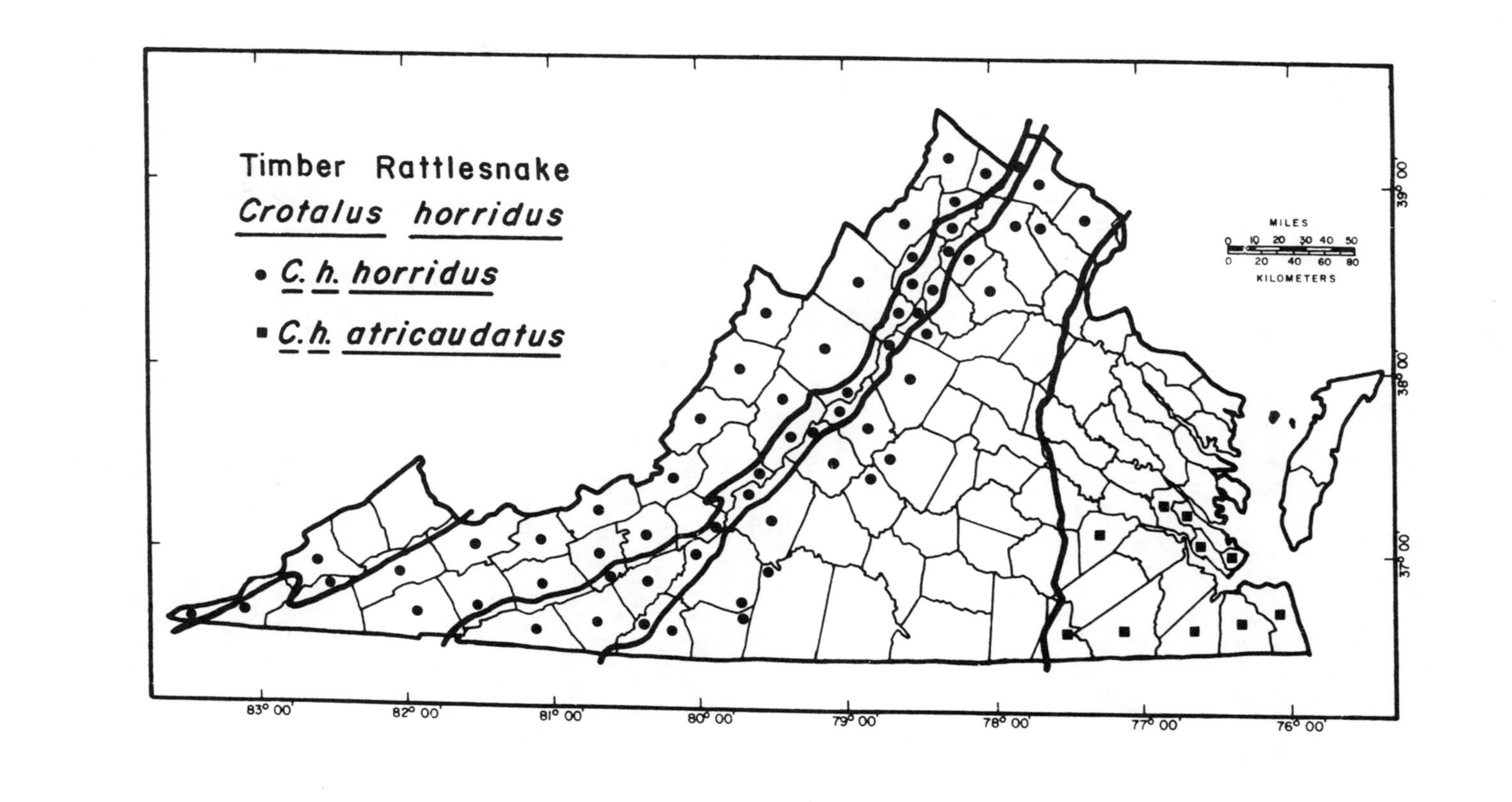
Timber Rattlesnake
Crotalus horridus
C. h. horridus
C. h. atricaudatus
MILES
0 10 20 30 40 50
0 20 40 60 80
KILOMETERS
39° 00'
38° 00'
37° 00'
83° 00'
82° 00'
81° 00'
80° 00'
79° 00'
78° 00'
77° 00'
76° 00'

Plates 51–52
Crotalus horridus horridus Linnaeus. Timber Rattlesnake. See description above. Western half of Virginia.

Plate 53
Crotalus horridus atricaudatus Latreille. Canebrake Rattlesnake. See description above. Southeastern Virginia.
Similar Species: Not likely to be confused with any of our other snakes.

Habitat. The northern subspecies is found mainly in hilly, rocky country. Mountainous areas, rocky ledges, and bluffs provide favorable habitat. Individuals also frequent briar patches, clearings, and blueberry patches during the summer months. Dens inhabited by timber rattlesnakes in Virginia have been reported at elevations ranging from 1700 to 4500 feet (Wood, 1954). The southern subspecies, on the other hand, inhabits river bottomlands, low pine woods, swamps, bayheads, and canebrakes. The name canebrake rattlesnake is derived from this snake's association with thickets of cane, a bamboolike plant of the southeastern bottomlands.

Range. This species ranges from New Hampshire, Vermont, New York, Ohio, Indiana, Illinois, Wisconsin, and Minnesota south to northern Florida and the Gulf Coast. The range extends westward to central Texas, central Oklahoma, eastern Kansas, and southeastern Nebraska. In Virginia, the timber rattlesnake inhabits the western portion of the state, while the canebrake rattlesnake is found only in the southeastern portion of the state. Russ (1973) stated: "It is probable the Dismal Swamp now holds the only significant habitat for this snake (canebrake rattlesnake) in Virginia since much of its historic range has been 'developed' by man."

The canebrake rattlesnake has officially been classified as a state endangered species in Virginia (Mitchell and Schwab, 1991).

Habits. Although rattlesnakes usually will make a buzzing sound with their rattle when disturbed, this may not always be the case. If they should suddenly be in danger, they will not hesitate to strike without rattling. The rattle consists of segments of dried skin. Newborn snakes have a thickened area at the tip of the tail known as the "pre-button." When the snakes are several days old, they shed their skin for the first time. The "pre-button" is also shed, exposing the "button," a bell-shaped, horny structure at the tip of the tail. Each time the snake sheds its skin during its lifetime, a remnant of the old skin remains and forms a new segment of the rattle. Depending upon the age, size, and physical condition of a snake, it may shed its skin from one to four or five times each year. Because of this variation and because the distalmost rattles often become broken and lost, these structures usually cannot be used to accurately determine the age of a rattlesnake.

Many rattlesnakes exhibit a unique behavioral phenomenon when in the presence of a kingsnake (Cowles, 1938). Instead of coiling and striking, a rattlesnake keeps its head and tail against the ground and raises the midsection of its body. As the kingsnake approaches, the elevated part of the rattlesnake's body is slammed down violently in an effort to club the attacker. The rattlesnake ordinarily does not bite until it has been encircled by the kingsnake's coils. Experiments by Bogert (1941) and Cowles and Phelan (1958) provide evidence that it is the odor of the kingsnake which elicits the fear response in rattlesnakes.

Reproduction. Timber rattlesnakes are ovoviviparous. Mating occurs in late summer with ovulation taking place in late May and early June of the following year (Martin, 1993). The mean age of probable first-time reproducers was 7.8 years (range 5–11 years). Martin found a mean interval of 3.08 years (range 2–4 years) between successive reproductions for marked females. As many as 15 young are born during late summer and fall. A 44-inch female collected on August 18 at the Mountain Lake Biological Station contained 9 fully developed eggs that showed practically no signs of embryonic development (Smyth, 1949). A study of the canebrake rattlesnake by Gibbons (1972) revealed that males are probably reproductively active in their fourth year, but that females do not have their first litter until they are approximately 6 years old. Females were found to apparently have a biennial reproductive cycle, thus giving birth to young every second year.

Newborn rattlesnakes are fully equipped with fangs and poison. Although the quantity of poison is considerably less than that in an adult, the young should be considered potentially dangerous.

Longevity. Maximum known age: 30 years, 2 months, 1 day (Bowler, 1977).

Food. Rattlesnakes feed primarily on small mammals such as mice, rats, rabbits, and squirrels. Reinert et al. (1984) described these snakes as sit and wait predators. The predominant ambush position consisted of coiling adjacent to a fallen log with the head positioned perpendicular to the log's long axis. The chin and/or a portion of the body often contacted the lateral surface of the log. From this position, snakes apparently ambushed small mammals that used the upper surfaces of logs as runways.

Uhler, Cottam, and Clarke (1939) analyzed the contents of 141 food-containing stomachs of this snake from the George Washington National Forest in Virginia. They found that mice made up 38% of the diet, squirrels and chipmunks comprised 25%, rabbits 18%, shrews 5%, and birds (chiefly songbirds) 13%. The stomach of one snake contained a bat. Examination of the stomachs of 12 timber rattlesnakes taken at the Mountain Lake Biological Station revealed an exclusively mammalian diet. Mammals identified included mice, chipmunks, and shrews. A wood rat had been previously recorded from a rattlesnake at the station (Smyth, 1949).

Enemies. Like the other pit vipers, rattlesnakes have few enemies. Man is the greatest threat to their existence, with most individuals killing any rattlesnake that happens to cross their path. A few rattlesnakes may also be taken by kingsnakes. A spotted skunk (*Spilogale putorius*) killed a 3-foot rattlesnake in Virginia (Anonymous, 1940*a*).

In Captivity. Captivity is not recommended. These snakes do not become docile in captivity. Many captives will not eat and will slowly become emaciated and starve to death if not released. Extreme caution must be exercised in feeding and cleaning the cage.

Folklore. The age of a rattlesnake, in most cases, cannot be determined by the number of rattles it possesses. A rattlesnake adds an additional rattle each time it sheds its skin, which may be three or four times a year. Furthermore, segments of the rattle are often broken off during the normal movements of the snake, especially if it is crawling over rough terrain. Most adult snakes will have between five and nine rattles.

Another belief about the rattlesnake is that you can tell its sex by its color. Although a rattlesnake's color has nothing to do with its sex, Netting (1932), Gloyd (1940), and Klauber (1956) did conclude that black individuals were generally males, whereas females were nearly always yellow. A study by Schaefer (1969) in Pennsylvania, however, found no significant differences in the frequency of occurrence of the color phases between the sexes.

Location of Specimens. AMNH, ART, CAS, CM, CWM, DU, DWL, FMNH, GMU, GMW, MLBS, NVCC, SDNHM, UI, UMMZ, USFWS, USNM, VCU, VPI and SU.

APPENDIX A

Checklist of Species and Subspecies of Snakes in Virginia

Family Colubridae

Nerodia taxispilota (Holbrook)	Brown water snake
Nerodia erythrogaster erythrogaster (Forster)	Red-belly water snake
Nerodia sipedon sipedon (Linnaeus)	Northern water snake
Nerodia sipedon pleuralis Cope	Midland water snake
Regina septemvittata (Say)	Queen snake
Regina rigida rigida (Say)	Glossy crayfish snake
Storeria dekayi dekayi (Holbrook)	Northern brown snake
Storeria dekayi wrightorum Trapido	Midland brown snake
Storeria occipitomaculata occipitomaculata (Storer)	Northern red-belly snake
Thamnophis sirtalis sirtalis (Linnaeus)	Eastern garter snake
Thamnophis sauritus sauritus (Linnaeus)	Eastern ribbon snake
Virginia valeriae valeriae Baird and Girard	Eastern smooth earth snake
Virginia valeriae pulchra Richmond	Mountain earth snake
Virginia striatula (Linnaeus)	Rough earth snake
Heterodon platirhinos Latreille	Eastern hognose snake
Diadophis punctatus edwardsi (Merrem)	Northern ringneck snake
Diadophis punctatus punctatus (Linnaeus)	Southern ringneck snake
Carphophis amoenus amoenus (Say)	Eastern worm snake
Farancia abacura abacura (Holbrook)	Eastern mud snake
Farancia erytrogramma erytrogramma (Latreille)	Rainbow snake
Coluber constrictor constrictor Linnaeus	Northern black racer
Opheodrys aestivus aestivus (Linnaeus)	Rough green snake
Opheodrys aestivus conanti Grobman	Barrier island rough green snake
Opheodrys vernalis vernalis (Harlan)	Eastern smooth green snake
Elaphe guttata guttata (Linnaeus)	Corn snake
Elaphe obsoleta obsoleta (Say)	Black rat snake

Pituophis melanoleucus melanoleucus (Daudin)	Northern pine snake
Lampropeltis getula getula (Linnaeus)	Eastern kingsnake
Lampropeltis getula niger (Yarrow)	Black kingsnake
Lampropeltis triangulum triangulum (Lacepede)	Eastern milk snake
Lampropeltis triangulum elapsoides (Holbrook)	Scarlet kingsnake
Lampropeltis calligaster rhombomaculata (Holbrook)	Mole kingsnake
Cemophora coccinea copei Jan	Northern scarlet snake
Tantilla coronata Baird and Girard	Southeastern crowned snake
Family Viperidae	
Agkistrodon contortrix mokasen Beauvois	Northern copperhead
Agkistrodon piscivorus piscivorus (Lacepede)	Eastern cottonmouth
Crotalus horridus horridus Linnaeus	Timber rattlesnake
Crotalus horridus atricaudatus Latreille	Canebrake rattlesnake

APPENDIX B

General References on Snakes

Allen, E. R., and W. T. Neill. 1954. Keep them alive! 2d edition. Ross Allen's Reptile Institute, Special Publ. no. 1. 26 pp.

Carr, A. 1963. The reptiles. Life Nature Library. Time, Inc., New York. 192 pp.

Cochran, D. M., and C. J. Goin. 1970. The new field book of reptiles and amphibians. G. P. Putnam's Sons, New York. 359 pp.

Conant, R., and J. T. Collins. 1991. A field guide to reptiles and amphibians of eastern and central North America. 3d edition. Houghton Mifflin Company, Boston. 450 pp.

Ernst, C. H., and R. W. Barbour. 1989. Snakes of eastern North America. George Mason Univ. Press, Fairfax, Virginia. 282 pp.

Goin, C. J., O. B. Goin, and G. R. Zug. 1978. Introduction to herpetology. 3d edition. W. H. Freeman and Co., San Francisco. 378 pp.

Gross, R. B. 1973. Snakes. Four Winds Press, New York. 63 pp.

Harrison, H. H. 1971. The world of the snake. J. B. Lippincott Co., Philadelphia. 160 pp.

Huntington, H. E. 1973. Let's look at reptiles. Doubleday and Company, Garden City, N.Y. 108 pp.

Kauffeld, C. F. 1969. Snakes: the keeper and the kept. Doubleday and Co., Garden City, N.Y. 249 pp.

Leviton, A. E. 1970. Reptiles and amphibians of North America. Doubleday and Co., Garden City, N.Y. 252 pp.

Oliver, J. A. 1955. The natural history of North American amphibians and reptiles. D. Van Nostrand Co., Princeton, N.J. 359 pp.

Pope, C. H. 1937. Snakes alive and how they live. Viking Press, New York. 238 pp.

Porter, K. R. 1972. Herpetology. W. B. Saunders Co., Philadelphia. 524 pp.

Schmidt, K. P., and R. F. Inger. 1957. Living reptiles of the world. Hanover House, Garden City, N.Y. 287 pp.

Smith, H. M. 1965. Snakes as pets. 3d edition. TFH Publications, Jersey City, N.J. 126 pp.

Smith, H., and H. S. Zim. 1987. Reptiles and amphibians. Golden Books, Western Publ., New York. 160 pp.
Wright, A. H., and A. A. Wright. 1957. Handbook of snakes of the United States and Canada. Comstock Publishing Associates, Ithaca, N.Y. Vol. I, 564 pp.; Vol. II, 541 pp.; Vol. III, 179 pp.

LITERATURE CITED AND BIBLIOGRAPHY

All references cited in the text are included here. In addition, an attempt has been made to compile as complete a bibliography as possible on the snakes of Virginia; all references preceded by an asterisk form this bibliography. If anyone knows of references that have been omitted, please inform the authors so that future revisions may be complete.

*Abbott, J. M. 1979. Crows versus black snake. Raven 50 (1):13.

*Addington, L. F. 1967. Hunting rattlers. Va. Wildl. 28 (7):20.

Aldridge, R. D. 1992. Oviductal anatomy and seasonal sperm storage in the southeastern crowned snake (*Tantilla coronata*). Copeia 1992 (4):1103–1106.

*Allard, H. A. 1945. A color variant of the eastern worm snake. Copeia 1945 (1):42.

*Alvord, C., and L. Bidgood. 1912. First explorations of the Trans-Allegheny regions by the Virginians, 1650–1674. Cleveland: Arthur H. Clark.

*Anderson, D. 1979. Stone Mountain. Va. Wildl. 40 (5):10–12.

*Andrews, J. D. 1971. Fish for beauty in Dismal Swamp! Va. J. Sci. 22 (1):5–13.

*Anonymous. 1940a. Spotted skunk battles rattlesnake. Va. Wildl. 4 (1):38.

*Anonymous. 1940b. Aquatic-minded turkeys. Va. Wildl. 4 (1):40.

*Anonymous. 1953a. Virginia's three poisonous snakes. Va. Wildl. 14 (5):27.

*Anonymous. 1953b. Snake collecting hobby of Colonial Heights lad. Va. Wildl. 14 (8):24.

*Anonymous. 1956a. Rare colorful snake. Va. Wildl. 17 (8):24.

*Anonymous. 1956b. The woodpecker and the snake. Va. Wildl. 17 (9):23.

*Anonymous. 1956c. The drumming log. Va. Wildl. 17 (10):25.

*Anonymous. 1957a. Captive rainbow snake is expectant mother. Va. Wildl. 18 (1):23.

*Anonymous. 1957b. 44-inch copperhead. Va. Wildl. 18 (9):26.

*Anonymous. 1959a. Best-of-month (April) find goes to society co-founder Witt now overseas. Va. Herp. Soc. Bull. no. 11:3.

*Anonymous. 1959b. The drumming log. Va. Wildl. 20 (9):25.

*Anonymous. 1960. The drumming log. Va. Wildl. 21 (10):22–23.

*Anonymous. 1961. Snaking for scouts. Va. Wildl. 22 (7):24.

*Anonymous. 1962. Notes on Virginian herpetology. Va. Herp. Soc. Bull. no. 29:6–7.

*Anonymous. 1964a. Fauquier County collecting notes. Va. Herp. Soc. Bull. no. 35:6.

*Anonymous. 1964b. Dismal Swamp collecting notes. Va. Herp. Soc. Bull. no. 36:7.

*Anonymous. 1965. Collection notes, Cape Henry, Va., Virginia Beach area. Va. Herp. Soc. Bull. no. 43:2–3.

*Anonymous. 1968. List of Virginia amphibians and reptiles. Va. Herp. Soc. Bull. no. 56:2–6.

*Anonymous. 1972. Dismal Swamp is natural landmark. The Virginia Outdoors 3 (1):6.

*Anonymous. 1973. The Great Dismal Swamp: a wild place for wildlife. The Virginia Outdoors 3 (3):6.

*Anonymous. 1974. The Dismal Swamp for all to enjoy. Va. Wildl. 35 (4):19–20, 24.

*Anonymous. 1977. Gap in Virginia range of smooth earth snake to be filled? Va. Herp. Soc. Bull. no. 82:4.

Anonymous. 1978. Note on red-sided garter snakes emerging from hibernation. Ecology USA 7 (4):28.

*Anonymous. Undated. A checklist of Virginia's mammals, birds, reptiles, and amphibians. Va. Comm. of Game and Inland Fisheries, Richmond, Va.

*Anonymous. Undated. The effect of man, weather, and food sources on raptors at Quantico and surrounding areas. Unpublished report, Quantico Marine Corps Base, Quantico, Virginia. 20 pp.

*Auffenberg, W. 1955. A reconsideration of the racer, *Coluber constrictor,* in eastern United States. Tulane Stud. Zool. 2:89–155.

*Ausband, S. 1990. Snakes that go bump in the night. Va. Wildl. 51 (1):4–7.

*Bader, R. N. 1983. Field notes: *Elaphe obsoleta obsoleta* (black rat snake). Catesbeiana 3 (2):13.

*Bader, R. N. 1984. The herpetofauna of South Isle Plantation. Catesbeiana 4 (1):3–4.

*Bailey, R. M. 1949. Temperature toleration of garter snakes in hibernation. Ecology 30:238–242.

*Baird, S. F., and C. Girard. 1853. Catalog of North American reptiles in the Museum of the Smithsonian Institution. Smithsonian Misc. Coll. vol. 2.

*Barbour, R. W. 1950. The reptiles of Big Black Mountain, Harlan County, Kentucky. Copeia 1950 (2):100–107.

*Barbour, R. W. 1971. Amphibians and reptiles of Kentucky. Lexington: University Press of Kentucky.

Barbour, R. W., M. J. Harvey, and J. W. Hardin. 1969. Home range, movements, and activity of the eastern worm snake, *Carphophis amoenus amoenus.* Ecology 50 (3):470–476.

*Barringer, P. B. 1892. The venomous reptiles of the United States, with the treatment of wounds inflicted by them. Trans. So. Surg. and Gyn. Assn. 1891, 4:283–300. Also Gaillard's Medical Journal 55 (1):7–19.

*Bartram, W. 1792. Travels through North and South Carolina, Georgia, east and west Florida. London. Reprinted by The Beehive Press, Savannah, Ga. 1973.

*Bazuin, J. B., Jr. 1983a. Reptiles and amphibians of the dioritic section of the green springs igneous intrusion, Louisa County, Virginia. Catesbeiana 3 (1):13–16.

*Bazuin, J. B., Jr. 1983b. Reptile and amphibian records from the Virginia Piedmont, 1975 to 1981. Catesbeiana 3 (2):3–6.

*Bazuin, J. B., Jr. 1987. Field notes: *Opheodrys aestivus* (rough green snake). Catesbeiana 7 (2):20.

*Behler, J. L., and F. W. King. 1979. The Audubon Society field guide to North American reptiles and amphibians. New York: Alfred A. Knopf.

*Belton, R. 1980. Copperhead. Va. Wildl. 41 (8):26–28.

*Beverley, R. 1705. The history and present state of Virginia, in four parts. Printed for R. Parker, London.

*Blanchard, F. N. 1920. Three new snakes of the genus *Lampropeltis*. Occ. Pap. Mus. Zool., Univ. Mich. 81:1–10.

*Blanchard, F. N. 1921. A revision of the king snakes; genus *Lampropeltis*. Bull. U.S. Nat. Mus. 114:1–260.

*Blanchard, F. N. 1923. The snakes of the genus *Virginia*. Pap. Mich. Acad. Sci., Arts and Letters 3:343–365.

*Blanchard, F. N. 1924. The status of *Amphiardis inornatus* (Garman). Copeia 1924: 83–85.

*Blanchard, F. N. 1925. The forms of *Carphophis*. Pap. Mich. Acad. Sci., Art and Letters 4 (1):527–530.

*Blanchard, F. N. 1942. The ring-necked snakes of the genus *Diadophis*. Bull. Chicago Acad. Sci. 7 (1):1–144.

*Blane, W. H. 1824. An excursion through the United States and Canada during the years 1822–23, by an English gentleman. London.

*Blaney, R. M. 1977. Systematics of the common kingsnake, *Lampropeltis getulus* (Linnaeus). Tulane Stud. Zool. and Bot. 19 (3–4):47–103.

*Blaney, R. M. 1979. *Lampropeltis calligaster*. Cat. Amer. Amphib. and Reptiles 229.1–229.2.

*Blem, C. R. 1978. The Virginia Commonwealth University herpetological collection. Va. Herp. Soc. Bull. no. 85:5.

*Blem, C. R. 1979. Predation of black rat snakes on a bank swallow colony. Wilson Bull. 91 (1):135–137.

*Blem, C. R. 1981a. Reproduction of the eastern cottonmouth *Agkistrodon piscivorus piscivorus* (Serpentes: Viperidae) at the northern edge of its range. Brimleyana 5:117–120.

*Blem, C. R. 1981b. *Heterodon platyrhinos*. Cat. Amer. Amphib. and Reptiles 282:1–2.

*Blem, C. R. 1982. Biennial reproduction in snakes: an alternative hypothesis. Copeia 1982:961–963.

*Blem, C. R., and K. L. Blem. 1990. Metabolic acclimation in three species of sympatric, semi-aquatic snakes. Comp. Biochem. Physiol. A Comp. Physiol. 97 (2):259–264.

*Blem, C. R., and L. B. Blem. 1985. Notes on *Virginia* (Reptilia: Colubridae) in Virginia. Brimleyana 11:87–95.

*Blem, C. R., and L. B. Blem. 1990. Lipid reserves of the brown water snake *Nerodia taxispilota*. Comp. Biochem. Physiol. A Comp. Physiol. 97 (3):367–372.

*Blem, C. R., and K. B. Killeen. 1993. Circadian metabolic cycles in eastern cottonmouths and brown water snakes. J. Herpetology 27 (3):341–344.

*Blem, C. R., and C. Roeding. 1983. Intergradation among ringneck snakes, *Diadophis punctatus* (Linnaeus) in Virginia. Va. J. Sci. 34 (4):207–214.

Bogert, C. M. 1941. Sensory cues used by rattlesnakes in their recognition of ophidian enemies. Ann. New York Acad. Sci. 41:329–343.

*Bonavita, J. 1982. The canebrake rattlesnake. Va. Wildl. 43 (6):16–17.

*Boo, T. 1974. Letters. Va. Wildl. 35 (8):3.

*Booker, K. A., and W. H. Yongue, Jr. 1979. Occurrence of Cytotoddia (=*Toddia:* Protozoa: Sporozoa) in *Nerodia* (=*Natrix*) *sipedon* (northern water snake) from an area in southwestern Virginia. Va. J. Sci. 30 (2):46. Abstract.

*Booker, K. A., and W. H. Yongue, Jr. 1982. Cytotoddia (=*Toddia*) infection of serpentes and its incidence in two geographical areas. Va. J. Sci. 33 (2):11–21.

Bothner, R. C. 1974. Some observations on the feeding habits of the cottonmouth in southeastern Georgia. J. Herpetology 8 (3):257–258.

Bowler, J. K. 1977. Longevity of reptiles and amphibians in North American collections. Soc. for the Study of Amphib. and Reptiles, Herp. Circ. no. 6.

*Boyd, W. K. 1929. William Byrd's histories of the dividing line betwixt Virginia and North Carolina. Raleigh: North Carolina Historical Commission.

*Brady, M. K. 1925. Notes on the herpetology of Hog Island. Copeia no. 137:110–111.

*Brady, M. K. 1927. Notes on the reptiles and amphibians of the Dismal Swamp. Copeia no. 162:26–29.

*Bragdon, D. E. 1953. A histochemical study of the lipids of the corpus luteum of pregnancy in the water snake, *Natrix sipedon.* Va. J. Sci. 4:273. Abstract.

Bragg, A. N. 1960. Is *Heterodon* venomous? Herpetologica 16:121–123.

*Brimley, C. S. 1918. Brief comparison of the herpetological faunas of North Carolina and Virginia. J. Elisha Mitchell Sci. Soc. 34:146–147.

*Brittle, D. L. 1969. Collected herpetofauna in Caroline County, Virginia. Va. J. Sci. 20 (3):110. Abstract.

*Brittle, D. L. 1969. Herpetofauna collected in Caroline County, Va. Va. Herp. Soc. Bull. no. 60:3–6.

*Brothers, D. R. 1992. An introduction to snakes of the Dismal Swamp region of North Carolina and Virginia. Boise, Idaho: Edgewood Probes, Inc.

*Brown, A. E. 1901. A review of the genera and species of American snakes, north of Mexico. Proc. Acad. Nat. Sci. Phila. 53:10–110.

*Brown, C. W. 1983. A study of variation in eastern timber rattlesnakes, *Crotalus horridus.* Va. J. Sci. 34 (3):117. Abstract.

*Brown, C. W., and C. H. Ernst. 1986. A study of variation in eastern timber rattlesnakes, *Crotalus horridus* Linnae (Serpentes: Viperidae). Brimleyana 12:57–74.

*Brown, W. S., and W. H. Martin. 1989. Geographic variation in female reproductive ecology of the timber rattlesnake, *Crotalus horridus.* Paper presented at First World Congress of Herpetology, Canterbury, England. September 11–19, 1989. Abstract: Catesbeiana 10 (2):48.

*Bruce, P. A. 1924. History of Virginia. Vol. 1. Colonial Period, 1607–1763. Chicago: American Historical Society.

*Bruenderman, S. 1993. Endangered species in Virginia. Focus on: Fish, amphibians, reptiles, and mollusks. Va. Wildl. 54 (2):10–16.

*Bruner, F. D. P. 1975. Southern living with rattlesnakes. Va. Wildl. 36 (5):20–21.

*Buhlmann, K. A. 1989. Field notes: *Crotalus horridus atricaudatus* (canebrake rattlesnake). Catesbeiana 9 (2):34.

*Buhlmann, K. A., and M. S. Hayslett. 1991. Herpetofauna of Chippokes Plantation State Park. Catesbeiana 11 (2):33–34.

*Buhlmann, K. A., A. H. Savitsky, B. A. Savitsky, and J. C. Mitchell. 1993. *Regina rigida* (Glossy crayfish snake). Herpetol. Review 24(4): 156–157.

*Bulmer, W. 1985. Report on an unbanded population of *Nerodia sipedon* from Virginia. Va. J. Sci. 36 (2):106. Abstract.

*Burch, P. R. 1940a. Snakes of the Allegheny Plateau in Virginia. Va. J. Sci. 1 (2–3):35–40.

*Burch, P. R. 1940b. Snakes of the Allegheny Plateau of Virginia. Va. Wildl. 4 (2):66–70.

Burger, J., R. T. Zappalorti, J. Dowdell, T. Georgiadis, J. Hill, and M. Gochfeld. 1992. Subterranean predation on pine snakes (*Pituophis melanoleucus*). J. Herpetology 26 (3):259–263.

*Burger, W. L. 1958. List of Virginian amphibians and reptiles. Va. Herp. Soc. Bull. Supplement to no. 4:1–5.

*Burger, W. L. 1958, revised 1959. A checklist of Virginian amphibians and reptiles. Virginia Fisheries Laboratory (now VIMS), Gloucester Point, Va. *In:* A checklist of Virginia's mammals, birds, fishes, reptiles, and amphibians. Reprinted from Virginia Wildlife, September, 1959.

*Burger, W. L. 1975. Herpetological specimens collected in Lee County, Virginia: (II) Reptiles. Va. Herp. Soc. Bull. no. 76:1–2, 6–7.

*Burkett, R. D. 1966. Natural history of cottonmouth moccasin, *Agkistrodon piscivorus* (Reptilia). Univ. Kansas Publ., Mus. Nat. Hist. 17 (9):435–491.

*Burnaby, A. 1798. Travels through the middle settlements in the North America in the years 1759 and 1760 with observations upon the state of the colonies. 3d edition. London: Mews-Gate. Reprinted by A. M. Kelley, New York. 1970.

*Bursey, C. R. 1986. Histological aspects of natural eustrongyloid infections of the northern water snake, *Nerodia sipedon.* J. Wildl. Disease 22 (4):527–532.

*Byrd, W. 1728. History of the dividing line. Pages 1–277 *in* Bassett, J. S. (ed.). 1901. The writings of Colonel William Byrd of Westover in Virginia, Esq. Burt Franklin, New York.

*Byrd, W. 1733. A journey to the land of Eden. Pages 279–329 *in* Bassett, J. S. (ed.). 1901. The writings of Colonel William Byrd of Westover in Virginia, Esq. Burt Franklin, New York.

Carpenter, C. C. 1952. Growth and maturity of the three species of *Thamnophis* in Michigan. Copeia 1952 (4):237–243.

Carpenter, C. C. 1953. A study of hibernacula and hibernating associations of snakes and amphibians in Michigan. Ecology 34 (1):74–80.

*Carroll, R. P. 1929. Survey of Cedar Creek in the Natural Bridge estate. Game and Fish Conservationist 9 (3):53–56.

*Carroll, R. P. 1950. Amphibia and reptiles. Pages 195–211 *in* The James River Basin: past, present, and future. Richmond: Va. Acad. Sci.

*Castiglioni, L. 1790. Viaggio negli Stati Uniti dell'America settentrionale fatto negli anni 1785, 1786, e 1787. Milan: Nella Stamperia di Guiseppe Marelli.

*Cerone, T. H. 1983. The natural history of *Virginia valeriae pulchra* (Serpentes: Colubridae). Ph.D. diss., St. Bonaventure Univ., St. Bonaventure. N.Y.

*Chasseur, T. 1875. The Nottoway region. Forest and Stream 5 (16):242.

*Christman, S. P. 1982. *Storeria dekayi.* Cat. Amer. Amphib. Rept. 306:1–4.

*Clark, D. R., Jr. 1968. A proposal of specific status for the western worm snake, *Carphophis amoenus vermis* (Kennicott). Herpetologica 24 (2):104–112.

Clark, D. R., Jr. 1970. Loss of left oviduct in the colubrid snake genus *Tantilla.* Herpetologica 26:130–133.

*Clarke, R. F. 1953. Alligator escapees in southeastern Virginia. Herpetologica 9:71–72.

*Clarke, S. 1670. Plantations of the English in America. London.

Clausen, H. J. 1936. Observations on the brown snake *Storeria dekayi* (Holbrook), with especial reference to the habits and birth of young. Copeia 1936 (2):98–102.

*Clayton, J. 1694. A continuation of Mr. John Clayton's account of Virginia. Philos. Trans. Royal Soc. London 18 (210):121–135.

*Cliburn, J. W. 1960. The phylogeny and zoogeography of North American *Natrix*. Ph.D. diss., Univ. of Alabama, Tuscaloosa.

*Clifford, M. 1973. Collecting notes: Amelia and Nottoway counties, Va. Va. Herp. Soc. Bull. no. 70:7–8.

*Clifford, M. J. 1975. Reptiles and amphibians. Extension Div. Publ. 676. Virginia Polytechnic Institute and State University.

*Clifford, M. J. 1976. Relative abundance and seasonal activity of snakes in Amelia County. Va. Herp. Soc. Bull. no. 79:4–6.

*Coale, C. B. 1878. The life and adventures of Wilburn Waters, the famous hunter and trapper of White Top Mountain; embracing early history of southwestern Virginia. Richmond: G. W. Gary and Co. Reprinted by Commonwealth Press, Radford, Va. 1976.

*Cochran, D. M. 1961. Type specimens of reptiles and amphibians in the United States National Museum. U.S. Natl. Mus. Bull. 220:1–291.

*Cochran, D. M., and C. J. Goin. 1970. The new field book of reptiles and amphibians. New York: G. P. Putnam's Sons.

*Coggin, J. L. 1955. Along forest waterways. Va. Wildl. 16 (9):16.

*Coleman, R. T. 1872. Snake story. Virginia Clin. Rec. 2 (4):137.

*Collins, J. T. 1966. Collecting in Caroline County. Va. Herp. Soc. Bull. no. 48:4–5.

Collins, J. T. 1990. Standard common and current scientific names for North American amphibians and reptiles. Soc. for the Study of Amphib. and Reptiles, Herp. Circular no. 19.

*Collins, J. T., and J. L. Knight. 1980. *Crotalus horridus*. Cat. Amer. Amphib. Rept. 253:1–2.

*Conant, R. 1943a. The milk snakes of the Atlantic Coastal Plain. Proc. New England Zool. Club. 22:3–24.

*Conant, R. 1943b. *Natrix erythrogaster erythrogaster* in the northeastern part of its range. Herpetologica 2 (5):83–86.

*Conant, R. 1945. An annotated check list of the amphibians and reptiles of the Del-Mar-Va Peninsula. Soc. Nat. Hist. Delaware.

*Conant, R. 1946. Intergradation among ring-necked snakes from southern New Jersey and the Del-Mar-Va Peninsula. Bull. Chicago Acad. Sci. 7 (10):473–482.

*Conant, R. 1948. Regeneration of clipped subcaudal scales in a pilot black snake. Nat. Hist. Misc. 13:1–2.

*Conant, R. 1949. Two new races of *Natrix erythrogaster*. Copeia 1949 (1):1–15.

*Conant, R. 1976. Reptiles and amphibians of the Virginia Barrier Islands, including several islands not examined in the present study. *In* Dueser, R. E., et al. The vertebrate zoogeography of the Virginia Coast Reserve, Table

31:5080509. From a report in the Virginia Coast Reserve study, G. J. Hennessey, Study Director. vol. 1. The ecosystem description.

*Conant, R. 1981. Herpetofauna (reptiles and amphibians) of the Virginia Coast Reserve. Pages 6–7 *in* Macleod, B., and G. J. Hennessey (eds.). The islands, official newsletter of the Virginia Coast Reserve, The Nature Conservancy, Arlington, Va.

*Conant, R. 1993. The Delmarva Peninsula. The Maryland Naturalist 37 (1–2):7–21.

*Conant, R., and J. T. Collins. 1991. A field guide to reptiles and amphibians of eastern and central North America. 3d edition. Boston: Houghton Mifflin Co.

*Conant, R., J. C. Mitchell, and C. A. Pague. 1990. Herpetofauna of the Virginia Barrier Islands. Va. J. Sci. 41 (4A):364–380.

*Conn, J. S. 1978. The hognose . . . a charmer. Va. Wildl. 39 (4):26.

*Conner, R. N., and I. D. Prather. 1976. Snakes: benign villains? Va. Wildl. 37 (10):22–23.

*Cooper, J. E. 1960. Collective notes on cave-associated vertebrates. Balto. Grotto News 3 (10):152–158.

*Cope, E. D. 1860. Catalogue of Colubridae in the Museum of the Academy of Natural Sciences of Philadelphia. I. Calamarinae. Proc. Acad. Nat. Sci. Phila.:74–79.

*Cope, E. D. 1875. Check-list of North American Batrachia and Reptilia; with a systematic list of the higher groups, and an essay on geographical distribution based on the specimens contained in the United States National Museum. Bull. U.S. Nat. Mus. 1:1–104.

*Cope, E. D. 1892. A critical review of the characters and variations of the snakes of North America. Proc. U.S. Nat. Mus. 14:589–694.

*Cope, E. D. 1895. A new locality for *Abastor erythrogrammus*. Amer. Nat. 29:588.

*Cope, E. D. 1896. The geographical distribution of Batrachia and Reptilia in North America. Amer. Nat. 30:886–902, 1003–1026.

*Cope, E. D. 1900. The crocodilians, lizards, and snakes of North America. Report of U.S. Nat. Mus. for 1898:153–1270.

Cowles, R. B. 1938. Unusual defense postures assumed by rattlesnakes. Copeia 1938 (1):13.

Cowles, R. B., and R. L. Phelan. 1958. Olfaction in rattlesnakes. Copeia 1958:77–83.

*Craig, C. 1967. Bedford County collecting notes. Va. Herp. Soc. Bull. no. 54:5.

*Craig, C. M. 1973. Letters. Hognose not rattler. Va. Wildl. 34 (3):3.

*Crawford, E. R., and R. A. S. Wright. 1990. Field notes: *Opheodrys aestivus* (rough green snake). Catesbeiana 10 (1):21.

Criddle, S. 1937. Snakes from an ant hill. Copeia 1937:142.

*Cross, G. H. 1974. Snakes and their control. MT no. 8G—Wildlife Coop. Extension Service, Virginia Polytechnic Institute and State University, Blacksburg.

*Croy, S. 1984. Field notes: *Lampropeltis getulus niger* (black kingsnake). Catesbeiana 4 (1):12.

*Davis, H. J. 1971. The Great Dismal Swamp. Its history, folklore, and science. Murfreesboro, N.C.: Johnson Publ. Co.

*Davis, N. S., Jr., and F. L. Rice. 1883. Descriptive catalog of North American Batrachia and Reptilia, found east of Mississippi River. Bull. Illinois State Lab. Nat. Hist. 5:1–66.

*DeKay, J. E. 1842. Zoology of New York. Part III. Reptiles and Amphibia. Natural History of New York, Albany.

*Delzell, D. E. 1971. The Dismal Swamp—its natural history. Living Wilderness, Winter 1970–71:29–33.

*Delzell, D. E. 1979. A provisional checklist of amphibians and reptiles in the Dismal Swamp area, with comments on their range of distribution. Pages 244–260 *in* Kirk, P. W., Jr. (ed.). The Great Dismal Swamp. Charlottesville: University Press of Virginia.

*de Rageot, R. H. 1957. Predation on small mammals in the Dismal Swamp, Virginia J. Mammal. 38 (2):281.

*de Rageot, R. H. 1964. Herpetofauna of Surry County, Virginia. Va. Herp. Soc. Bull. no. 40:3–6.

Dietrich, R. V. 1970. V = F (S . . .). Pages 67–99 *in* Holt, P. C. (ed.). The distributional history of the biota of the Southern Appalachians. Virginia Polytechnic Institute and State University Research Div. Monograph 2.

*Ditmars, R. L. 1907. The reptile book. Garden City, N.Y.: Doubleday, Page and Co.

*Ditmars, R. L. 1931. Snakes of the world. New York: Macmillan Company.

Ditmars, R. L. 1935. Serpents of the northeastern states. New York Zoological Society.

*Ditmars, R. L. 1936. The reptiles of North America. Garden City, N.Y.: Doubleday, Doran and Co.

*Ditmars, R. L. 1944. Reptiles of the world. New York: Macmillan Co.

*Dolan, D. 1990. Marines discover rattlesnake at OCS. Quantico Sentry [newspaper], Oct. 19, pp. A6–A7.

*Drowne, T. P. 1900. A trip to Fauquier Co., Virginia; with notes on the specimens obtained. The Museum 6 (3):38–45.

Dudderar, G. R., and J. R. Beck (eds.). 1973. Rats and their control. Virginia Polytechnic Institute and State University Extension Div. Publ. 272:1–12.

*Duellman, W. E. 1949. An unusual habitat for the keeled green snake. Herpetologica 5:144.

*Dueser, R. D., W. C. Brown, S. A. McCuskey, and G. S. Hogue. 1976. Vertebrate zoogeography of the Virginia Coast Reserve. Pages 439–518 *in* Virginia Coast Reserve Study, Ecosystem description. The Nature Conservancy, Arlington, Va.

*Dundee, H. A., and J. Ewan. 1979. An overlooked description of a United States snake from the eighteenth century writing of Luigi Castiglioni. J. Herpetology 13 (2):216–217.

Dunlap, K. D., and J. W. Lang. 1990. Offspring sex ratio varies with maternal size in the common garter snake, *Thamnophis sirtalis*. Copeia 1990 (2):568–570.

*Dunn, E. R. 1915a. List of amphibians and reptiles observed in the summers of 1912, 1913, and 1914, in Nelson County, Virginia. Copeia no. 18:5–7.

*Dunn, E. R. 1915b. Number of young produced by certain snakes. Copeia no. 22:37.

*Dunn, E. R. 1915c. List of reptiles and amphibians from Clark County, Va. Copeia no. 25:62–63.

*Dunn, E. R. 1915d. The variations of a brood of watersnakes. Proc. Biol. Soc. Wash. 28:61–68.

*Dunn, E. R. 1917. The pine snake in Virginia. Copeia no. 51:101.

*Dunn, E. R. 1918. A preliminary list of the reptiles and amphibians of Virginia. Copeia no. 53:16–27.

*Dunn, E. R. 1919. *Tantilla coronata* in Virginia. Copeia no. 76:100.

*Dunn, E. R. 1920. Some reptiles and amphibians from Virginia, North Carolina, Tennessee, and Alabama. Proc. Biol. Soc. Wash. 33:129–137.

*Dunn, E. R. 1936. List of Virginia amphibians and reptiles. Mimeographed. Haverford, Pa.

*Dunn, E. R., and G. C. Wood. 1939. Notes on eastern snakes of the genus *Coluber*. Notulae Naturae 5:1–4.

*Eckerlin, R. P. 1991. The herpetofauna of George Washington Birthplace National Monument, Virginia. Catesbeiana 11 (1):11–17.

*Edgren, R. A. 1955. The natural history of the hognosed snakes, genus *Heterodon:* a review. Herpetologica 11 (2):105–115.

*Edmond, C. 1939. Snakes on trial. Va. Wildl. 3 (2):6.

*Edwards, S. R. (ed.). 1975. Collections of preserved amphibians and reptiles in the United States. Soc. for the Study of Amphib. and Reptiles, Misc. Publ. Herp. Circ. no. 3.

*Elkins, P. 1974. Virginia's loneliest river. Va. Wildl. 35 (9):7–9.

*Engeling, G. 1969a. Collecting notes: Hampton–Newport News, Va. Va. Herp. Soc. Bull. no. 61:5.

*Engeling, G. A. 1969b. Reptiles and amphibians of York Co., Va., and the Newport News–Hampton area. Va. Herp. Soc. Bull. no. 62:1–3.

*Ernst, C. H., and R. W. Barbour. 1989. Snakes of eastern North America. Fairfax, Virginia: George Mason University Press.

*Ernst, C. H., S. W. Gotte, and J. E. Lovich. 1985. Reproduction in the mole kingsnake, *Lampropeltis calligaster rhombomaculata*. Bull. Maryland Herp. Soc. 21 (1):16–22.

*Ernst, C. H., and A. F. Laemmerzahl. 1989. Eastern hognose snake eats spotted salamander. Bull. Maryland Herp. Soc. 25 (1):25–26.

*Ernst, S. G. 1963. Copperheads in suburbia (I). Va. Herp. Soc. Bull. no. 34:1–2.

*Evans, A. 1934. Arese, Francesco. A trip to the prairies and in the interior of North America [1837–1838]. Trans. A. Evans. New York: Harbor Press.

*Ewan, J., and N. Ewan. 1970. John Banister and his natural history of Virginia, 1678–1692. Urbana: University of Illinois Press.

*Farr, R. 1979. Letters. Helpful snakes. Va. Wildl. 40 (8):3.

*Fay, L. P. 1986. Wisconsinan herpetofaunas of the central Appalachians. Pages 126–128 *in* McDonald, J. N., and S. O. Bird (eds.). The Quaternary of Virginia–a symposium volume. Virginia Div. of Mineral Resources Publ. 75.

Fenneman, N. M. 1938. Physiography of eastern United States. New York: McGraw Hill Book Co.

Finneran, L. C. 1953. Aggregation behavior of the female copperhead, *Agkistrodon contortrix mokeson,* during gestation. Copeia 1953 (1):61–62.

*Fitch, H. S. 1960. Autecology of the copperhead. Univ. of Kansas Publ. Mus. Nat. Hist. 13:85–288.

*Fitch, H. S. 1963. Natural history of the racer *Coluber constrictor.* Univ. Kansas Publ. Mus. Nat. Hist. 15 (8):351–468.

*Fitch, H. S. 1965. An ecological study of the garter snake, *Thamnophis sirtalis.* Univ. Kansas Publ. Mus. Nat. Hist. 15 (10):493–564.

Fitch, H. S. 1970. Reproductive cycles of lizards and snakes. Univ. Kansas Mus. Nat. Hist., Misc. Publ. no. 52:1–247.

*Fitch, H. S. 1980. *Thamnophis sirtalis.* Cat. Amer. Amphib. Rept. 270:1–4.

*Forbes, J. E., and F. B. Leftwich. 1967. Respiratory activities of the vibratory muscles of *Crotalus horridus, Agkistrodon contortrix* and *Thamnophis sirtalis.* Va. J. Sci. 18 (4):159. Abstract.

*Fowler, H. W. 1925. Records of amphibians and reptiles for Delaware, Maryland, and Virginia. III. Virginia. Copeia no. 146:65–67.

*Fowler, J. A. 1945. Notes on *Cemophora coccinea* (Blumenbach) in Maryland and the District of Columbia vicinity. Proc. Biol. Soc. Wash. 58:89–90.

*Fowler, J. A. 1970. Notes on *Cemophora coccinea* (Blumenbach) in Maryland and the District of Columbia vicinity. Bull. Maryland Herp. Soc. 6 (4):86–87.

Fraker, M. A. 1970. Home range and homing in the watersnake, *Natrix sipedon sipedon.* Copeia 1970 (4):665–673.

*Freer, R. S., and F. T. Hanenkrat. 1980. The central Blue Ridge. Part III: The fauna of the Blue Ridge. Va. Wildl. 41 (9):16–20.

*Gagnon, R. J. 1978. Earliest reported 1978 Va. collection record goes to R. J. Gagnon and rescue squad. Va. Herp. Soc. Bull. no. 86:4.

*Garman, S. 1883. The reptiles and batrachians of North America. Part I. Ophidia.—Serpents. Mus. Comp. Zool., Harvard Univ., Memoir 8 (3):1–185.

*Garman, S. 1884. The North American reptiles and batrachians. Bull. Essex Instit. 16 (1–3):3–46.

Gibbons, J. W. 1972. Reproduction, growth, and sexual dimorphism in the canebrake rattlesnake (*Crotalus horridus atricaudatus*). Copeia 1972 (2):222–226.

*Gibbons, J. W., and J. W. Coker. 1978. Herpetofaunal colonization patterns of Atlantic coast barrier islands. Amer. Midl. Nat. 99:219–233.

*Gibbons, J. W., J. W. Coker, and T. M. Murphy, Jr. 1977. Selected aspects of the life history of the rainbow snake (*Farancia erytrogramma*). Herpetologica 33 (3):276–281.

*Gillam, H. 1962. Third southeastern crowned snake found in Virginia. Va. Wildl. 23 (8):25.

*Gilley, S. 1983. The black rat snake. Va. Wildl. 44 (6):22.

Gloyd, H. K. 1934. Studies on the breeding habits and young of the copperhead, *Agkistrodon mokasen* Beauvois. Pap. Mich. Acad. Sci., Arts, and Letters 19:587–604.

*Gloyd, H. K. 1940. The rattlesnakes, genera *Sistrurus* and *Crotalus.* A study in zoogeography and evolution. Chicago Acad. Sci. Spec. Publ. no. 4.

*Gloyd, H. K. 1947. Notes on the courtship and mating behavior of certain snakes. Nat. Hist. Misc. 12:1–4.

*Gloyd, H. K., and R. Conant. 1943. A synopsis of the American forms of *Agkistrodon* (copperheads and moccasins). Bull. Chicago Acad. Sci. 7 (2):147–170.

*Gloyd, H. K., and R. Conant (eds.). 1989. Snakes of the *Agkistrodon* complex: a monographic review. Soc. for the Study of Amphib. and Reptiles, Contrib. to Herpetology no. 6.

*Goodrich, S. G. 1859. Illustrated natural history of the animal kingdom. . . . Vol. 2. New York: Derby and Jackson.

*Goodwin, O. K., and J. T. Wood. 1956. Distribution of poisonous snakes on the York-James Peninsula: a zoogeographic mystery. Va. J. Sci. 7:17–21.

Gosner, K. L. 1942. Lip curling of the red-bellied snake. Copeia 1942 (3):181–182.

*Grobman, A. B. 1984. Scutellation variation in *Opheodrys aestivus*. Bull. Florida St. Mus. 29 (4):153–170.

*Grobman, A. B. 1992. Metamerism in the snake *Opheodrys vernalis,* with a description of a new subspecies. J. Herpetology 26 (2):175–186.

Grogan, W. L., Jr. 1974. Effects of accidental envenomation from the saliva of the eastern hognose snake, *Heterodon platyrhinos*. Herpetologica 30 (3):248–249.

*Groves, F. 1978. A case of twinning in the ringneck snake, *Diadophis punctatus edwardsi*. Bull. Maryland Herp. Soc. 14 (1):48–49.

Groves, J. D., and P. S. Sachs. 1973. Eggs and young of the scarlet king snake, *Lampropeltis triangulum elapsoides*. J. Herpetology 7 (4):389–390.

*Guilday, J. E. 1962. The Pleistocene local fauna of the Natural Chimneys, Augusta County, Virginia. Ann. Carnegie Mus. 36:87–122.

*Guillaudeu, R. L. 1963. First aid treatment for poisonous snakebite. Va. Herp. Soc. Bull. no. 34:1.

*Hardy, J. D., Jr. 1972. Reptiles of the Chesapeake Bay region. Chesapeake Sci. 13 (suppl.):S128–S134.

*Hardy, J. D., Jr., and J. E. Olmon. 1971. Zoogeography in action: observations on the movements of terrestrial vertebrates across open water. Maryland Herp. Soc. Newsletter 1971 (1):4–6.

Harlow, P., and G. Grigg. 1984. Shivering thermogenesis in a brooding diamond python, *Python spilotes spilotes*. Copeia 1984 (4):959–965.

*Harrison, G. H. 1959. The timber rattlesnake. Va. Wildl. 20 (5):25.

*Hay, W. P. 1902. A list of the batrachians and reptiles of the District of Columbia and vicinity. Proc. Biol. Soc. Wash. 15:121–145.

*Hayslett, M. S. 1992. Narrative of the 1992 VHS spring meeting. Catesbeiana 12 (2):36–43.

*Hayslett, M. S. 1992. Field notes: *Coluber constrictor constrictor* (northern black racer). Catesbeiana 12 (2):46–47.

*Hayslett, M. S. 1993. Field notes: *Opheodrys aestivus* (rough green snake). Catesbeiana 13 (1):11.

*Heiner, A. B. 1967. The creek named Sarah. Va. Wildl. 28 (4):10–11.

*Hensley, M. 1959. Albinism in North American amphibians and reptiles. Publ. Mus., Mich. State Univ., Biological Series 1 (4):133–159.

*Hill, J. M., and T. A. Pierson. 1986. The herpetofauna of Caledon State Park, Virginia. Catesbeiana 6 (1):11–17.

*Hoback, W. W., and T. W. Green. 1953. Treatment of snake venom poisoning with cortisone and corticotropin. J. Amer. Med. Assoc. 152 (3):236–237.

Hoffman, L. H. 1970. Observations on gestation in the garter snake, *Thamnophis sirtalis sirtalis*. Copeia 1970 (4):779–780.

*Hoffman, R. L. 1945. Notes on the herpetological fauna of Alleghany County, Virginia. Herpetologica 2 (7–8):199–205.

*Hoffman, R. L. 1953. Interesting herpesian records from Camp Pickett, Virginia. Herpetologica 8 (4):171–174.

*Hoffman, R. L. 1969. The biotic regions of Virginia. Virginia Polytechnic Institute and State University Research Div. Bull. 48:23–62.

*Hoffman, R. L. 1977. Scarlet snake record for western Virginia. Va. Herp. Soc. Bull. no. 83:3.

*Hoffman, R. L. 1985. The herpetofauna of Alleghany County, Virginia. Catesbeiana 5 (1):3–12.

*Hoffman, R. L. 1986. The herpetofauna of Alleghany County, Virginia. Part 3. Class Reptilia. Catesbeiana 6 (1):4–10.

*Hoffman, R. L. 1987. The herpetofauna of Alleghany County, Virginia. Part 4. Biogeographic inferences. Catesbeiana 7 (1):5–14.

*Hoffman, R. L. 1988. Field notes: *Lampropeltis getulus niger* (black kingsnake). Catesbeiana 8 (2):32.

*Hoffman, R. L., and J. C. Mitchell. 1994. Paul R. Burch's herpetological collection at Radford College, Virginia: a valuable resource lost. Catesbeiana 14 (1):3–12.

*Holbrook, J. E. 1842. North American herpetology; or, a description of the reptiles inhabiting the United States. 2d edition. Philadelphia.

*Holder, J. B. 1885. Animate creation; popular edition of "Our living world," a natural history by the Rev. J. G. Wood. Revised and adapted to American zoology by Joseph B. Holder. New York: Selmar Hess.

*Holman, J. A., and J. N. McDonald. 1986. A late Quaternary herpetofauna from Saltville, Virginia. Brimleyana 12:85–100.

*Hopley, C. C. 1882. Snakes: curiosities and wonders of serpent life. London: Griffith and Farran.

*Huheey, J. E. 1959. Distribution and variation in the glossy water snake, *Natrix rigida* (Say). Copeia 1959 (4):303–311.

*Humphries, B. 1975. Letters. Black snake attacks, retreats. Va. Wildl. 36 (8):3.

*Hunter, A. 1876. The Dismal Swamp. Forest and Stream 5 (23):353–354.

*Hutchinson, R. H. 1929. On the incidence of snake-bite poisoning in the United States and the results of newer methods of treatment. Bull. Antivenin Instit. of America 3:43–57.

*Hutchinson, R. H. 1930. Further notes on the incidence of snake-bite poisoning in the United States. Bull. Antivenin Instit. of America 4:40–43.

*Hutchison, V. H. 1956. An annotated list of the amphibians and reptiles of Giles County, Virginia. Va. J. Sci. 7 (2):80–86.

*Jackson, J. J. 1983. Snakes of the southeastern United States. Georgia Extension Service.

*Jopson, H. G. M. 1971. The origin of the reptile fauna of the southern Appalachians. Pages 189–196 *in* Holt, P. C. (ed.). The distributional history of the biota of the Southern Appalachians. Virginia Polytechnic Institute and State University Research Div. Monograph 4.

*Jopson, H. G. M. 1984. Amphibians and reptiles from Rockingham County, Virginia. Catesbeiana 4 (2):3–9.

*Kauffeld, C. F. 1957. Snakes and snake hunting. Garden City, N.Y.: Hanover House.

*Kellner, W. C. 1955. The Dismal–No Business area of Giles and Bland counties. Va. Wildl. 16 (8):10–12.

*Kercheval, S. 1925. A history of the Valley of Virginia. 4th ed. Strasburg, Va.: Shenandoah Publishing House.

*Klauber, L. M. 1972. Rattlesnakes. Their habits, life histories, and influence on mankind. 2d ed. 2 volumes. Berkeley: University of California Press.

*Klimkiewicz, M. K. 1970. Autumn migration flyway one—Mason Neck. Atlantic Naturalist 25 (4):160–164.

*Klimkiewicz, M. K. 1972. Reptiles of Mason Neck. Atlantic Naturalist 27 (1):20–25.

*Kruk, D. R. 1969. Letters. A big one. Va. Wildl. 30 (4):3.

Lachner, E. A. 1942. An aggregation of snakes and salamanders during hibernation. Copeia 1942 (4):262–263.

*Lane, M. 1986. Field notes: *Lampropeltis t. triangulum x L. t. elapsoides* (Coastal Plains milksnake). Catesbeiana 6 (2):14.

*Leavitt, D. A. 1978. Dismal Swamp, report of a 1977 YCC participant. Va. Herp. Soc. Bull. no. 86:3–4.

*Lederer, J. 1672. The discoveries of John Lederer, in three several marches from Virginia, to the west of Carolina, and other parts of the continent: begun in March 1669, and ended in September 1670. Together with a general map of the whole territory which he traversed. Collected and translated out of Latin from his discourse and writings, by Sir William Talbot, Baronet. Printed by J. C. for S. Heyrich, London, 1672; reprinted for G. P. Humphrey, Rochester, N.Y., 1902.

*Lee, D. S. 1972. List of the amphibians and reptiles of Assateague Island. Bull. Maryland Herp. Soc. 8 (4):90–95.

*Leftwich, F. B., G. C. Schaefer, J. E. Turner, and J. E. Forbes. 1967. Respiration in the vibratory muscle of *Crotalus horridus* (Reptilia: Crotalidae). Va. J. Sci. 18 (4):161. Abstract.

*Leviton, A. E. 1970. Reptiles and amphibians of North America. Garden City, N.Y.: Doubleday and Co.

*Lewis, J. B. 1928. Wildlife of the Dismal Swamp. Game and Fish Conservationist 8 (4):95–98.

*Lewis, J. B. 1940. Mammals of Amelia County, Virginia. J. Mammal. 21(4):422–428.

*Linzey, D. W. 1959. Further records of the smooth green snake in the Virginia Blue Ridge mountains. Herpetologica 15 (2):94.

*Linzey, D. W. (ed.). 1980. Proceedings of the Symposium on Endangered and Threatened Plants and Animals of Virginia. Blacksburg: Virginia Polytechnic Institute and State University.

*Llewellyn, L. M. 1943. The common pine snake in West Virginia. Copeia 1943 (2):129.

*Lough, M. W. 1978. Blue birds always lose! Va. Wildl. 39 (1):22.

*Luckeydoo, A. K., and C. R. Blem. 1993. High temperature triggers circadian activity rhythms of brown water snakes (*Nerodia taxispilota*). Va. J. Sci. 44 (2):111. Abstract.

*Lynn, W. G. 1936. Reptile records from Stafford County, Virginia. Copeia 1936 (3):169–171.

McAlister, W. H. 1963. Evidence of mild toxicity in the saliva of the hognose snake (*Heterodon*). Herpetologica 19:132–137.

*McCauley, R. H., Jr. 1939. An extension of the range of *Abastor erythrogrammus*. Copeia 1939 (1):54.

*McCauley, R. H., Jr. 1941. A redescription of *Lampropeltis triangulum temporalis* (Cope). Copeia 1941 (3):146–150.

*McCranie, J. R. 1983. *Nerodia taxispilota*. Cat. Amer. Amphib. Rept. 331:1–2.

*McDaniel, V. R., and J. P. Karges. 1983. *Farancia abacura*. Cat. Amer. Amphib. Rept. 314:1–2.

*Marshall, H. G. 1972. Lake Drummond—heart of Great Dismal Swamp. Atlantic Naturalist 27 (2):60–64.

*Martin, J. R. 1965. Letter to the editor. Va. Herp. Soc. Bull. no. 41:4.

*Martin, J. R., and J. T. Wood. 1955. Notes on the poisonous snakes of the Dismal Swamp area. Herpetologica 11 (3):237–238.

Martin, W. F., and R. B. Huey. 1971. The function of the epiglottis in sound production (hissing) of *Pituophis melanoleucus*. Copeia 1971 (4):752–754.

*Martin, W. H. 1986 (1987). Reproduction of the timber rattlesnake in northwestern Virginia. Paper presented at annual meeting of Soc. for the Study of Amphib. and Reptiles/Herpetologists' League, Springfield, Mo. August 1986. Abstract in Catesbeiana 7 (1):32–33. 1987.

*Martin, W. H. 1988. Life history of the timber rattlesnake. Catesbeiana 8 (1):9–12.

*Martin, W. H. 1989. The timber rattlesnake, *Crotalus horridus,* in the Appalachian Mountains of eastern North America. Paper presented at First World Congress of Herpetology, Canterbury, England. September 11–19, 1989. Abstract in Catesbeiana 10 (2):49.

*Martin, W. H. 1992. Phenology of the timber rattlesnake (*Crotalus horridus*) in an unglaciated section of the Appalachian Mountains. Pages 259–277 *in* Campbell, J. A., and E. D. Brodie, Jr. (eds.). Biology of the pitvipers. Tyler, Texas: Selva.

*Martin, W. H. 1993. Reproduction of the timber rattlesnake (*Crotalus horridus*) in the Appalachian Mountains. J. Herpetology 27 (2):133–143.

*Martin, W. H., J. C. Mitchell, and R. Hoggard. 1992. *Crotalus horridus* (timber rattlesnake). Herpetological Review 23 (3):91.

*Martin, W. H., III. 1964. The timber rattlesnake on Virginia's upper Piedmont. Va. Herp. Soc. Bull. no. 40:1.

*Martin, W. H., III. 1976. Reptiles observed on the Skyline Drive and Blue Ridge Parkway, Va. Va. Herp. Soc. Bull. no. 81:1–3.

*Martin, W. H., III. 1983. Unpublished investigator's annual report, submitted to Shenandoah National Park Superintendent.

*Martin, W. H. 1984. Geographic distribution. *Lampropeltis calligaster rhombomaculata* (mole kingsnake). Herp. Review 15 (1):21.

*Martof, B. S., W. M. Palmer, J. R. Bailey, and J. R. Harrison III. 1980. Amphibians and reptiles of the Carolinas and Virginia. Chapel Hill: University of North Carolina Press.

Meade, G. P. 1940. Maternal care of eggs by *Farancia*. Herpetologica 2 (1):15–20.

*Meanley, B. 1971. The Dismal Swamp—its flora and fauna. Living Wilderness 34:34–37.

*Meanley, B. 1972. Swamps, river bottoms, and canebrakes. Barre, Mass.: Barre Publishers.

*Meanley, B. 1973. The Great Dismal Swamp. Washington, D.C.: Audubon Nat. Soc. Central Atlantic states.

*Medden, R. V. 1931. Tales of the rattlesnakes from the works of early travelers in America. Bull. Amer. Antivenin Instit. 5:24–27.

*Megonigal, J. P. 1985. Field notes: *Agkistrodon contortrix mokasen* (northern copperhead) and *Lampropeltis getulus getulus* (eastern kingsnake). Catesbeiana 5 (1):16.

*Merkle, D. 1987. Field notes: New county records for the queen snake *Regina septemvittata* in the central Piedmont of Virginia. Catesbeiana 7 (2): 19–20.

*Merkle, D. A. 1982. Genetic variation in the cottonmouth water moccasin *Agkistrodon piscivorous* at the northern edge of its distribution. Catesbeiana 2 (2):11. Abstract.

*Merkle, D. A. 1985. Genetic variation in the eastern cottonmouth, *Agkistrodon piscivorus piscivorus* (Lacepede) (Reptilia: Crotalidae) at the northern edge of its range. Brimleyana 11:55–61.

*Miller, G. S., Jr. 1902. A fully adult specimen of *Ophibolus rhombomaculatus*. Proc. Biol. Soc. Wash. 15:36.

*Mitchell, J. 1990. Eek, it's a snake. Va. Wildl. 51 (4):7–12.

*Mitchell, J. C. 1973. Geographic distribution: *Elaphe guttata guttata* (corn snake). HISS News—J. 1 (5):153.

*Mitchell, J. C. 1974a. Distribution of the corn snake in Virginia. Va. Herp. Soc. Bull. no. 74:3–5.

*Mitchell, J. C. 1974b. Notes on a cottonmouth from Petersburg, Virginia. Va. Herp. Soc. Bull. no. 75:5.

*Mitchell, J. C. 1974c. The snakes of Virginia. Part I. The poisonous snakes and their look alikes. Va. Wildl. 35 (2):16–18.

*Mitchell, J. C. 1974d. The snakes of Virginia. Part II. The harmless snakes that benefit man. Va. Wildl. 35 (4):12–13.

*Mitchell, J. C. 1974e. *Natrix taxispilota* (brown water snake). Herpetological Review 5 (3):70.

*Mitchell, J. C. 1976. Notes on reproduction in *Storeria dekayi* and *Virginia striatula* from Virginia and North Carolina. Bull. Maryland Herp. Soc. 12 (4):133–135.

*Mitchell, J. C. 1977. Geographic variation of *Elaphe guttata* (Reptilia: Serpentes) in the Atlantic Coastal Plain. Copeia 1977 (1):33–41.

*Mitchell, J. C. 1980. Viper's brood. A guide to identifying some of Virginia's juvenile snakes. Va. Wildl. 41 (9):8–10.

*Mitchell, J. C. 1981. Notes on male combat in two Virginia snakes, *Agkistrodon contortrix* and *Elaphe obsoleta*. Catesbeiana 1 (1):7–9.

*Mitchell, J. C. 1981. A bibliography of Virginia amphibians and reptiles. Smithsonian Herpetological Information Service no. 50.

*Mitchell, J. C. 1982a. Snake lore: fact or fiction. Va. Wildl. 43 (5):14–15.

*Mitchell, J. C. 1982b. *Farancia*. Cat. Amer. Amphib. and Reptiles 292:1–2.

*Mitchell, J. C. 1982c. *Farancia erytrogramma*. Cat. Amer. Amphib. and Reptiles 293:1–2.

*Mitchell, J. C. 1986. Amphibians and reptiles collected and observed on the Spring VHS meeting field trip in Caroline County, Virginia. Catesbeiana 6 (2):15–16.

*Mitchell, J. C. 1988a. Proposed list of amphibians and reptiles of special concern in Virginia. Catesbeiana 8 (2):29–30.

*Mitchell, J. C. 1988b. Russian herpetologists visit Virginia. Catesbeiana 8 (2):35.

*Mitchell, J. C. 1989. Proposed list of amphibians and reptiles of special concern in Virginia. Catesbeiana 9 (2):37–38.

*Mitchell, J. C. 1990a. Contributions to the history of Virginia herpetology. II. John B. Lewis' "List of reptiles observed in Amelia, Brunswick, and Norfolk Counties." Catesbeiana 10 (2):36–42.

*Mitchell, J. C. 1991a. Snakes—predators of amazing skill and grace. Va. Wildl. 52 (6):17–22.

*Mitchell, J. C. 1991b. Mountain earth snake, *Virginia valeriae pulchra*. Pages 461–462 *in* Terwilliger, K. (coordinator). Virginia's endangered species. Blacksburg: McDonald and Woodward Publishing Co.

*Mitchell, J. C. 1992a. Exotic delights—pleasures or plagues? Va. Wildl. 53 (6):4–8.

*Mitchell, J. C. 1992b. A snake of a different color. Va. Wildl. 53 (8):27–32.

*Mitchell, J. C. 1992c. The glossy crayfish snake. Va. Wildl. 53 (3):32–33.

*Mitchell, J. C. 1993a. Unearthing treasures: the quest to find out more about Virginia's amphibians and reptiles. Va. Wildl. 54 (2):18–22.

*Mitchell, J. C. 1993b. A muddy view of life. . . . Va. Wildl. 54 (2):22–23.

*Mitchell, J. C. 1993c. Tales of couch potatoes and heat-seeking missiles. Va. Wildl. 54 (6):9–15.

*Mitchell, J. C. 1994. Herptiles across the Commonwealth. Virginia Explorer 10 (1):14–16.

*Mitchell, J. C. 1994. Reptiles redeemed. Va. Wildl. 55(7):8–13.

*Mitchell, J. C. 1994. The reptiles of Virginia. Blue Ridge Summit, Pa.: Smithsonian Institution Press.

*Mitchell, J. C., and R. A. Beck. 1991. Free-ranging domestic cat predation on native vertebrates in rural and urban Virginia. Va. J. Sci. 42 (2):174. Abstract.

*Mitchell, J. C., and R. A. Beck. 1992. Free-ranging domestic cat predation on native vertebrates in rural and urban Virginia. Va. J. Sci. 43 (1B):198–207.

*Mitchell, J. C., R. Conant, and C. A. Pague. 1989. Herpetofauna of the Virginia Barrier Islands: distribution and biogeography. Paper presented at First World Congress of Herpetology, Canterbury, England. September 11–19, 1989. Abstract in Catesbeiana 10 (2):52–53.

*Mitchell, J. C., and W. H. Martin, III. 1981. Where the snakes are. Va. Wildl. 42 (6):8–9.

*Mitchell, J. C., and W. H. Mitchell. 1989. A preliminary survey of the amphibians and reptiles of Sweet Briar College, Virginia. Catesbeiana 9 (2):25–31.

*Mitchell, J. C., and C. A. Pague. 1981. Amphibians and reptiles of Virginia project: comments and update. Va. J. Sci. 32(3):95. Abstract.

*Mitchell, J. C., and C. A. Pague. 1984. Reptiles and amphibians of far southwestern Virginia: report on a biogeographical and ecological survey. Catesbeiana 4 (2):12–17.

*Mitchell, J. C., and C. A. Pague. 1985. Reptiles and amphibians of Virginia book: update 1985. Va. J. Sci. 36 (2):113. Abstract.

*Mitchell, J. C., and C. A. Pague. 1987. A review of reptiles of special concern in Virginia. Va. J. Sci. 38 (2):84. Abstract.

*Mitchell, J. C., and C. A. Pague. 1987. A review of reptiles of special concern in Virginia. Va. J. Sci. 38 (4):319–328.

*Mitchell, J. C., and C. A. Pague. 1988. Herpetofauna of the Virginia barrier islands. Va. J. Sci. 39 (2):119. Abstract.

*Mitchell, J. C., C. A. Pague, and D. L. Early. 1982. Life history: *Elaphe obsoleta.* Autophagy. Herpetological Review 13 (2):47.

*Mitchell, J. C., and D. Schwab. 1991. Canebrake rattlesnake (*Crotalus horridus atricaudatus)*. Pages 462–464 *in* Terwilliger, K. (coor.). Virginia's endangered species. Blacksburg: McDonald and Woodward Publishing Co.

*Mitchell, J. C., and G. R. Zug. 1984. Spermatogenic cycle of *Nerodia taxispilota* (Serpentes: Colubridae) in southcentral Virginia. Herpetologica 40 (2):200–204.

*Mitchell, J. W. 1985. They're the pits. Va. Wildl. 46 (6):27–29.

*Monson, G. 1971. Spring phenology. Atlantic Naturalist 26 (2):65–70.

*Moore, D. M. 1986. Mite infestation on snakes. Catesbeiana 6 (2):4–6.

*Morgan, M. 1953. Boy Scouts know their snakes. Va. Wildl. 14 (7):8–9, 22, 24.

*Mosby, H. S. 1948. Virginia's poisonous snakes and their venom. Va. Wildl. 9 (7):16–18.

*Mosby, H. S. 1964. Letters. Oddity. Va. Wildl. 25 (1):26.

*Mosby, H. S. 1972. A "wonder" drug from deer? Va. Wildl. 33 (9):17–18.

*Murray, E. 1970. Mountain Lake biological station. Va. Wildl. 31 (12):17–18.

*Murray, R. L. 1969. Collecting notes: Prince William County. Va. Herp. Soc. Bull. no. 61:5.

*Musick, J. A. 1972. Herptiles of the Maryland and Virginia coastal plain. Pages 213–242 *in* Wass, M. L. (ed.). A check list of the biota of lower Chesapeake Bay. Spec. Sci. Report no. 65, Va. Instit. Marine Sci., Gloucester Point, Va.

*Myers. E. 1973. Look-alike snakes. Va. Wildl. 34 (7):19–20.

*Neale, G. 1978. Black snake. Va. Wildl. 39 (8):22–23.

Neill, W. T. 1948a. Hibernation of amphibians and reptiles in Richmond County, Georgia. Herpetologica 4:107–114.

Neill, W. T. 1948b. The yellow tail of juvenile copperheads. Herpetologica 4 (5):161.

Neill, W. T. 1958. The occurrence of amphibians and reptiles in saltwater areas, and a bibliography. Mar. Sci. Gulf Carib. Bull. 8 (1):1–97.

Neill, W. T. 1960. The caudal lure of various juvenile snakes. Quart. J. Fla. Acad. Sci. 23 (3):173–200.

*Neill, W. T. 1964. Taxonomy, natural history, and zoogeography of the rainbow snake, *Farancia erythrogramma* (Palisot de Beauvois). Amer. Midl. Nat. 71 (2):257–295.

*Nelson, A. L. 1933. A preliminary report on the winter food of Virginia foxes. J. Mammal. 14 (1):40–43.

Netting, M. G. 1932. The poisonous snakes of Pennsylvania. Carnegie Mus. Vert. Zool. Pamphlet no. 1.

*Netting, M. G. 1936. The chain snake, *Lampropeltis getulus getulus* (L)., in West Virginia and Pennsylvania. Annals Carnegie Mus., Pittsburgh 25:77–82.

*Nicoletto, P. 1985. Some reptiles from Sinking Creek and Gap Mountain, Montgomery County, Virginia, April–June, 1983. Catesbeiana 5 (1):13–15.

*Norden, A. 1971. A corn snake *Elaphe guttata guttata,* from western Maryland. Maryland Herp. Soc. Bull. 7 (1):25–27.

*Obaugh, W. 1969. Blacksnake in slow motion. Va. Wildl. 30 (7):17.

Oliver, J. A. 1955. The natural history of North American amphibians and reptiles. Princeton, N.J.: D. Van Nostrand Co.

*Orrick, N. B. 1961. Letters. Helped survey Mt. Rogers. Va. Wildl. 22 (8):3.

*Ortenburger, A. I. 1928. The whip snakes and racers: genera *Masticophis* and *Coluber.* Mem. Univ. Mich. Mus. 1:1–247.

*Padgett, T. M. 1987. Field notes: Observations of courtship behavior in *Elaphe obsoleta* (black rat snake). Catesbeiana 7 (1):27.

*Pague, C. 1981. The captive maintenance of Virginia's watersnakes. Catesbeiana 1 (1):15–18.

*Pague, C. A. 1986. Hidden motion. Va. Wildl. 47 (8):13–15.

*Pague, C. A., and J. C. Mitchell. 1981. A herpetofaunal survey of Back Bay National Wildlife Refuge: a preliminary report. Va. J. Sci. 32 (3):96. Abstract.

*Pague, C. A., and J. C. Mitchell. 1982. A checklist of amphibians and reptiles of Back Bay National Wildlife Refuge and False Cape State Park, Virginia Beach, Virginia. Catesbeiana 2 (2):13–15.

*Pague, C. A., J. C. Mitchell, D. A. Young, and K. A. Buhlmann. 1990. Species composition and seasonal surface activity of terrestrial vertebrates in five northern Virginia Piedmont natural communities. Va. J. Sci. 41 (2):58. Abstract.

*Palmer, W. M. 1959. A second record of the glossy water snake in North Carolina. Herpetologica 15:47.

*Palmer, W. M. 1971. Distribution and variation of the Carolina pigmy rattlesnake, *Sistrurus miliarius miliarius* Linnaeus, in North Carolina. J. Herpetology 5 (1–2):39–44.

Palmer, W. M., and G. Tregembo. 1970. Notes on the natural history of the scarlet snake *Cemophora coccinea copei* Jan in North Carolina. Herpetologica 26 (3):300–303.

Parrish, H. M. 1957. Morality from snakebites, United States, 1950–54. Public Health Reports 72 (11):1027–1030.

*Parrish, H. M. 1963. Analysis of 460 fatalities from venomous animals in the United States. Amer. Jour. Med. Sci. 245 (2):35–47.

*Parrish, H. M. 1966. Incidence of treated snakebites in the United States. Public Health Reports 81 (3):269–276.

*Patrick, B. L., and W. Wieland. 1993. A survey of amphibians and reptiles on a proposed site for the Chesapeake Bay National Estuarine Research Reserve system in King George County, Virginia. Va. J. Sci. 44 (2):112. Abstract.

*Paul, J. R. 1967. Intergradation among ring-necked snakes in southeastern United States. J. Elisha Mitchell Sci. Soc. 83:98–102.

*Pilcher, A. 1972. Youth afield. Va. Wildl. 33 (10):25.

*Pisani, G. R., and J. T. Collins. 1971. The smooth earth snake, *Virginia valeriae* (Baird and Girard), in Kentucky. Trans. Kentucky Acad. Sci. 32:16–25.

*Pisani, G. R., J. T. Collins, and S. R. Edwards. 1973. A re-evaluation of the subspecies of *Crotalus horridus.* Trans. Kans. Acad. Sci. 75 (3):255–263.

*Platt, D. R. 1969. Natural history of the hognose snakes *Heterodon platyrhinos* and *Heterodon nasicus*. Univ. Kans. Pub. Mus. Nat. Hist. 18 (4):253–420.

*Plummer, M. V. 1987. Geographic variation in body size of green snakes (*Opheodrys aestivus*). Copeia 1987 (2):483–485.

Plummer, M. V. 1990. High predation on green snakes, *Opheodrys aestivus*. J. Herpetology 24 (3):327–328.

*Rae, S. 1974. The corn snake in Prince William County, Va. Va. Herp. Soc. Bull. no. 74:5–6.

*Rageot, R. H. 1957. Predation on small animals in the Dismal Swamp, Virginia. J. Mammal. 38 (2):281.

*Rageot, R. H. 1959. Awakening of the Swampland. Va. Herp. Soc. Bull. no. 10:2.

*Reed, C. F. 1956a. Contributions to the herpetology of Maryland and Delmarva. No. 5. Bibliography to the herpetology of Maryland, Delmarva, and the District of Columbia. Mimeo.

*Reed, C. F. 1956b. Contributions to the herpetology of Maryland and Delmarva. No. 8. An annotated check list of the snakes of Maryland and Delmarva. Mimeo.

*Reed, C. F. 1956c. Contributions to the herpetology of Maryland and Delmarva. No. 11. An annotated herpetofauna of the Del-Mar-Va Peninsula, including many new or additional localities. Mimeo.

*Reed, C. F. 1957a. Contributions to the herpetofauna of Virginia, 2: The reptiles and amphibians of Northern Neck. J. Wash. Acad. Sci. 47 (1):21–23.

*Reed, C. F. 1957b. Contributions to the herpetology of Virginia, 3: The herpetofauna of Accomac and Northampton counties, Va. J. Wash. Acad. Sci. 47 (3):89–91.

*Reed, C. F. 1958a. Contributions to the herpetology of Maryland and Delmarva. No. 13. Piedmont herpetofauna on coastal Delmarva. J. Wash. Acad. Sci. 48 (3):95–99.

*Reed, C. F. 1958b. Contributions to the herpetology of Maryland and Delmarva. No. 17. Southeastern herptiles with northern limits on coastal Maryland, Delmarva, and New Jersey. J. Wash. Acad. Sci. 48 (1):28–32.

Reinert, H. K., D. Cundall, and L. M. Bushar. 1984. Foraging behavior of the timber rattlesnake, *Crotalus horridus*. Copeia 1984 (4):976–981.

*Richmond, J. L. 1965. Notes on the snakes at Ashland, Hanover County, Virginia. Va. Herp. Soc. Bull. no. 41:2.

*Richmond, N. D. 1940. *Natrix rigida* Say in Virginia. Herpetologica 2 (1):21.

*Richmond, N. D. 1944. How *Natrix taxispilota* eats the channel catfish. Copeia 1944 (4):254.

*Richmond, N. D. 1945. The habits of the rainbow snake in Virginia. Copeia 1945 (1):28–30.

*Richmond, N. D. 1952. *Opheodrys aestivus* in aquatic habitats in Virginia. Herpetologica 8 (3):38.

*Richmond, N. D. 1954. Variation and sexual dimorphism in hatchlings of the rainbow snake, *Abastor erythrogrammus*. Copeia 1954 (2):87–92.

*Richmond, N. D. 1956. Autumn mating of the rough green snake. Herpetologica 12 (4):325.

*Richmond, N. D., and C. J. Goin. 1938. Notes on a collection of amphibians and reptiles from New Kent County, Virginia. Ann. Carnegie Mus. 27: 301–310.

Riemer, W. J. 1957. The snake *Farancia abacura:* an attended nest. Herpetologica 13 (1):31–32.

*Roble, S. M., and C. S. Hobson. 1994. Field notes. *Farancia erytrogramma* (Rainbow Snake). Catesbeiana 14 (1):15–16.

*Robertson, L. D. 1985. Life history notes. *Nerodia sipedon sipedon* (northern water snake). Herp. Review 16 (4):111.

*Rochefoucault, F. A. F., Liancourt, duc de. 1800. Travels through the United States of North America, etc., in the years 1795, 1796, and 1797. 2d ed. London.

*Roper, L. J. 1951. For a healthy vacation. Va. Wildl. 12:5–7, 12.

*Rossman, D. A. 1963a. The colubrid snake genus *Thamnophis:* a revision of the *sauritus* group. Bull. Florida State Mus. 7 (3):99–178.

Rossman, D. A. 1963b. Relationship and taxonomic status of the North American natricine snake genera *Liodytes, Regina,* and *Clonophis*. Occ. Pap. Mus. Zool. Louisiana State Univ. 29:1–29.

*Rossman, D. A. 1970. *Thamnophis sauritus* (Linnaeus). Cat. Amer. Amphib. and Reptiles 99:1–99.2.

Rossman, D. A., and W. G. Eberle. 1977. Partition of the genus *Natrix,* with preliminary observations on evolutionary trends in natricine snakes. Herpetologica 33 (1):34–43.

*Russ, W. P. 1973. The rare and endangered terrestrial vertebrates of Virginia. MS thesis, Virginia Polytechnic Institute and State University, Blacksburg.

*Russ, W. P. 1974. Endangered vertebrates of Virginia. Va. Wildl. 35 (9):13–15, 18.

*Ruthven, A. G. 1908. Variations and genetic relationships of the garter-snakes. Bull. U.S. Nat. Mus. 61:1–198.

*Salmon, T. 1736–1738. Modern history: or, the present state of all nations. Describing their respective situations, persons, habits, buildings, manners, laws and customs, religion and policy, arts and sciences, trades, manufactures and husbandry, plants, animals and minerals. Pages 339–374. London. Printed for the author.

Sanders, J. S., and J. S. Jacob. 1981. Thermal ecology of the copperhead (*Agkistrodon contortrix*). Herpetologica 37 (4):264–270.

*Sattler, P. W. 1990. Field notes: *Storeria o. occipitomaculata* (northern red-bellied snake). Catesbeiana 10 (2):45.

*Saylor, L. W. 1938. Hairy-tailed mole in Virginia. J. Mammal. 19 (2):247.

Schaefer, G. C. 1968. Sex independent ground color in the timber rattlesnake, *Crotalus horridus horridus*. Va. J. Sci. 19 (3):182. Abstract.

Schaefer, G. C. 1969. Sex independent ground color in the timber rattlesnake, *Crotalus horridus horridus*. Herpetologica 25 (1):65–66.

*Schatti, B., and L. D. Wilson. 1986. *Coluber.* Cat. Amer. Amphib. and Rept. 399:1–4.

*Schmidt, K. P. 1953. A check list of North American amphibians and reptiles. 6th edition. Amer. Soc. Ichthy. and Herp. Chicago: Univ. of Chicago Press.

*Schmidt, K. P., and R. F. Inger. 1957. Living reptiles of the world. Garden City, N.Y.: Hanover House.

Schuett, G. W., and J. C. Gillingham. 1988. Courtship and mating of the copperhead, *Agkistrodon contortrix*. Copeia 1988 (2):374–381.

*Schwaner, T. D., and J. M. Anderson. 1991. Geographic distribution. *Opheodrys aestivus* (rough green snake). Herpetological Review 22 (2):68.

*Schwab. D. 1985a. Tongue worms (Pentastomida) in a northern copperhead (*Agkistrodon contortrix mokasen*) (Daudin) from the Great Dismal Swamp, Virginia. Catesbeiana 5 (2):14.

*Schwab, D. 1985b. Field notes: *Coluber constrictor constrictor* (northern black racer). Catesbeiana 5 (2):15.

*Schwab, D. 1986a. Field notes: *Heterodon platyrhinos* (eastern hognose snake). Catesbeiana 6 (1):21.

*Schwab, D. 1986b. Field notes: *Heterodon platyrhinos* (eastern hognose snake). Catesbeiana 6 (2):14.

*Schwab, D. 1987a. Field notes: *Elaphe obsoleta obsoleta* (black rat snake). Catesbeiana 7 (1):27.

*Schwab, D. 1987b. Field notes: *Crotalus horridus* (timber rattlesnake). Catesbeiana 7 (2):20.

*Schwab, D. 1988a. Reptiles and amphibians observed in the Ware Creek watershed, James City County, Virginia. Catesbeiana 8 (1):3–7.

*Schwab, D. 1988b. Field notes: *Heterodon platyrhinos* (eastern hognose snake). Catesbeiana 8 (1):13.

*Schwab, D. 1988c. Field notes: *Coluber constrictor constrictor* (northern black racer). Catesbeiana 8 (2):32.

*Schwab, D. 1988d. Growth and rattle development in a captive timber rattlesnake, *Crotalus horridus*. Bull. Chicago Herp. Soc. 23:26–27.

*Scott, D. 1986. Notes on the eastern hognose snake, *Heterodon platyrhinos* Latreille (Squamata: Colubridae), on a Virginia barrier island. Brimleyana 12:51–55.

*Shaw, G. 1802. General Zoology or Systematic Natural History. Vol. 3, pt. II. Amphibia. London: G. Kearsley.

*Shelton, N. 1975. The nature of Shenandoah. Natural History Series, National Park Service. U.S. Dept. of the Interior, Washington, D.C.

Silver, J. 1928. Pilot black-snake feeding on the big brown bat. J. Mammal. 9 (2):149.

*Skinner, H. E., Jr. 1978. Sanctuary near the crowds. Va. Wildl. 39 (7):10–12.

*Sloane, H. 1734. Conjectures on the charming or fascinating power attributed to the rattle-snake: grounded on credible accounts, experiments, and observations. Phil. Trans. 38 (433):321–331.

*Smith, H. M. 1938. A review of the snake genus *Farancia*. Copeia 1938 (3):110–117.

Smith, H. M., and J. E. Huheey. 1960. The water snake genus *Regina*. Trans. Kansas Acad. Sci. 63:156–164.

*Smith, J. 1624. The generall historie of Virginia, New-England, and the Summer Isles. London: Printed by I. D. and I. H. for Michael Sparkes.

*Smyth, T. 1949. Notes on the timber rattlesnake at Mountain Lake, Va. Copeia 1949 (1):78.

*Solomon, G. B. 1974. Probable role of the timber rattlesnake, *Crotalus horridus*, in the release of *Capillaria hepatica* (Nematoda) eggs from small mammals. Va. J. Sci. 25 (4):182–184.

*Southall, L. 1965. Collecting notes on Chesterfield and Dinwiddie counties, Va. Va. Herp. Soc. Bull. no. 42:4.

*Spooner, J. J. 1793. Prince George County, Va. Mass. Hist. Soc. Coll. 3:86. Boston, 1910.

*Stahl, S. J., and J. C. Mitchell. 1985a. *Agkistrodon contortrix* in Virginia. Va. J. Sci. 36 (2):115. Abstract.

*Stahl, S. J., and J. C. Mitchell. 1985b. Reproductive cycle of the northern copperhead, *Agkistrodon contortrix,* in Virginia. Va. J. Sci. 36 (2):115. Abstract.

*Stansbury, C. F. 1925. The lake of the Great Dismal. New York: Albert and Charles Boni.

*Steirly, C. C. 1963. Eastern cottonmouth taken in Surry County, Va. Va. Herp. Soc. Bull. no. 34:4.

*Stejneger, L. H. 1891. Notes on some North American snakes. Proc. U.S. Nat. Mus. 14:501–505.

*Stejneger, L. 1895. The poisonous snakes of North America. Report of U.S. National Museum for 1893:337–487.

*Stejneger, L., and T. Barbour. 1943. A check list of North American amphibians and reptiles. Bull. Mus. Comp. Zool., Harvard Univ. 93 (1):1–260.

*Stickel, W. H. 1952. Venomous snakes of the United States and treatment of their bites. U.S. Fish and Wildlife Service, Wildlife Leaflet no. 399.

*Strachey, W. 1849. The historie of travaile into Virginia Brittannia by William Strachey, Gent. London: Hakluyt Society.

*Stull, O. G. 1940. Variations and relationships in the snakes of the genus *Pituophis.* U.S. Nat. Mus. Bull. 175:1–225.

*Taylor, E. A. 1949. Not all snakes are bad. Va. Wildl. 10 (7):10–11.

*Taylor, E. A. 1952. Some facts on Virginia poisonous snakes. Va. Wildl. 13 (5):18–19, 22.

*Taylor, E. A. 1956. Snakes and snake venom. Va. Wildl. 17 (7):22–23.

*Taylor, E. A. 1958. Virginia poisonous snakes. Va. Wildl. 19 (7):8–9.

*Taylor, J. W. 1975a. Scarlet king snake. Va. Wildl. 36 (1):10.

*Taylor, J. W. 1975b. The northern pine snake. Va. Wildl. 36 (3):27.

*Taylor, J. W. 1975c. Canebrake rattlesnake. Va. Wildl. 36 (4):21.

*Telford, S. R., Jr. 1966. Variation among the southeastern crowned snakes, genus *Tantilla.* Bull. Florida State Mus. 10 (7):261–304.

*Telford, S. R., Jr. 1982. *Tantilla coronata.* Cat. Amer. Amphib. Rept. 308:1–2.

*Thomas, B. 1976. The swamp. New York: W. W. Norton and Co.

*Thornton, W. M. 1896. Spottswood's expedition of 1716. Nat. Geog. Mag. 7 (8):265–269.

*Tobey, F., Jr. 1957. Harmless and often helpful snakes. Va. Wildl. 18 (4):8–9, 24.

*Tobey, F., Jr. 1960. The occurrence of poisonous snakes in Virginia, Maryland, and D. C. Va. Herp. Soc. Bull. no. 19:1–2.

*Tobey, F., Jr. 1961a. The southeastern crowned snake. Va. Wildl. 22 (5):8–9.

*Tobey, F., Jr. 1961b. The southeastern crowned snake—small, shy, secretive and smooth-scaled. Va. Herp. Soc. Bull. no. 24:1–4.

*Tobey, F., Jr. 1962. Report on the occurrence of poisonous snakes of Virginia, Maryland, and the District of Columbia. Va. Herp. Soc. Bull. no. 29: 1–4.

*Tobey, F. J. 1973. Great Dismal Swamp. Va. Herp. Soc. Bull. no. 71:1–2.

*Tobey, F. J. 1979. Amphibians and reptiles. Pages 375–414 *in* Linzey, D. W. (ed.). Proceedings of the symposium on endangered and threatened plants and animals of Virginia. Blacksburg: Virginia Polytechnic Institute and State University.

*Tobey, F. J. 1985. Virginia's amphibians and reptiles: a distributional survey. Purcellville, Va.: Virginia Herpetological Society.

*Tobey, F. J. 1986. Some amphibian and reptilian records from Loudoun Heights, Loudoun County, Virginia. Catesbeiana 6 (2):7–10.

*Tobey, F. J. 1989. Field notes: *Coluber constrictor constrictor* (northern black racer). Catesbeiana 9 (2):35.

*Tobey, F. J., Jr. 1963. Copperheads in suburbia (II). Va. Herp. Soc. Bull. no. 34:2–3.

*Tobey, F. J., Jr. 1964a. Chesterfield County collecting notes. Va. Herp. Soc. Bull. no. 35:7.

*Tobey, F. J., Jr. 1964b. An aid to identification of the snakes of Virginia. Va. Herp. Soc. Bull. no. 37:1–14.

*Tobey, F. J., Jr. 1969. Letters, comments, and ideas. Va. Herp. Soc. Bull. no. 60:7.

*Tobey, F. J., Jr. 1975. Letters. Out of range. Va. Wildl. 36 (8):3.

*Traister, J. 1979. Snakebite! Va. Wildl. 40 (8):13.

*Trapido, A. 1944. The snakes of the genus *Storeria*. Amer. Midl. Nat. 31: 1–84.

*Troubetzkoy, U. 1967. Seasons in Dismal Swamp—spring. Va. Wildl. 28 (6):17–20.

*Troubetzkoy, U. 1976. A look at Virginia—1607. Va. Wildl. 37 (7):9–13.

*Tuck, R. G., Jr. 1969. Collecting note, Dickenson Co. Va. Herp. Soc. Bull. no. 60:10.

*Tuck, R. G., Jr., M. K. Klimkiewicz, and K. C. Ferris. 1971. Notes on pilot blacksnake (*Elaphe obsoleta obsoleta*) (Serpentes: Colubridae) eggs and hatchlings. Bull. Maryland Herp. Soc. 7 (4):96–99.

*Tupacz, E. G. 1985. Field notes: *Nerodia erythrogaster* (redbelly water snake) and *Buteo lineatus* (red-shouldered hawk). Catesbeiana 5 (2):15.

*Tuttle, H. J. 1946. The snakes. Va. Wildl. 7 (6):14–15, 22.

*Uhler, F. M., C. Cottam, and T. E. Clarke. 1939. Food of snakes of the George Washington National Forest, Virginia. Trans. 4th North Amer. Wildl. Conf.: 605–622.

*Watson, S. 1953. Headwaters of the Potomac. Va. Wildl. 14 (11):10–12.

Weatherhead, P. J., and I. C. Robertson. 1990. Homing to food by black rat snakes (*Elaphe obsoleta*). Copeia 1990 (4):1164–1165.

*Wells, K. 1967. The northern red-bellied snake in Fairfax County. Va. Herp. Soc. Bull. no. 53:2.

*Werler, J. E., and J. McCallion. 1951. Notes on a collection of reptiles and amphibians from Princess Anne County, Virginia. Amer. Midl. Nat. 45 (1):245–252.

Wever, E. G., and J. A. Vernon. 1960. The problem of hearing in snakes. J. Auditory Research 1:77–83.

Wharton, C. H. 1960. Birth and behavior of a brood of cottonmouths, *Agkistrodon piscivorus piscivorus*, with notes on tail-luring. Herpetologica 16 (2):125–129.

*White, D. R., J. C. Mitchell, and W. S. Woolcott. 1982. Reproductive cycle and embryonic development of *Nerodia taxispilota* (Serpentes: Colubridae) at the northeastern edge of its range. Copeia 1982 (3):646–652.

*White, M. 1978. Wildlife kaleidoscope, copperhead caper. Va. Wildl. 39 (3):22–23.

*Wilbur, H. M. 1964. Collecting notes—Augusta and Rockbridge counties, Virginia. Va. Herp. Soc. Bull. no. 40:7–8.

*Williams, K. L. 1978. Systematics and natural history of the American milk snake, *Lampropeltis triangulum*. Milwaukee Public Museum Publ. in Biology and Geology no. 2:1–258.

*Williams, K. L. 1985. *Cemophora, C. coccinea*. Cat. Amer. Amphib. Rept. 374:1–4.

*Williams, K. L., and L. D. Wilson. 1967. A review of the colubrid snake genus *Cemophora* Cope. Tulane Stud. Zool. 13 (4):103–124.

*Williamson, G. M. 1964. Letter to the editor. Va. Herp. Soc. Bull. no. 40:8.

*Williamson, G. M. 1979. Canebrake rattlesnake account. Pages 407–409 *in* Tobey, F. J. Amphibians and reptiles. *In* Linzey, D. W. (ed.). Proceedings of the symposium on endangered and threatened plants and animals of Virginia. Blacksburg: Virginia Polytechnic Institute and State University.

Willson, P. 1908. Snake poisoning in the United States: a study based on an analysis of 740 cases. Arch. Int. Med. 1 (5):516–570.

*Wilson, L. D. 1978. *Coluber constrictor* Linnaeus. Cat. Amer. Amphib. and Reptiles 218.1–218.4.

*Wilson, L. D. 1982. *Tantilla*. Cat. Amer. Amphib. Rept. 307:1–4.

*Wilson, L. W., and S. Friddle. 1946. Notes on the king snake in West Virginia. Copeia 1946 (1):47–48.

*Witt, W. L. 1958. The smooth green snake in the Virginia Blue Ridge mountains. Herpetologica 14 (3):140.

*Witt, W. L. 1961a. List of Virginia amphibians and reptiles—which have been reported for three counties or less. Va. Herp. Soc. Bull. no. 24:5–6.

*Witt, W. L. 1961b. Plenty of "frontier" to Virginian herpetology. Va. Herp. Soc. Bull. no. 26:1–2.

*Witt, W. L. 1961c. Virginia collecting notes. Va. Herp. Soc. Bull. no. 27:7–8.

*Witt, W. L. 1962a. Problems in Virginian herpetology. Va. Herp. Soc. Bull. no. 28:3–4.

*Witt, W. L. 1962b. Reptiles and amphibians observed during statewide meeting. Va. Herp. Soc. Bull. no. 31:3–4.

*Witt, W. L. 1963a. Collection turned in to U.S. National Museum. Va. Herp. Soc. Bull. no. 34:6, 8.

*Witt, W. L. 1963b. Charles City County collecting. Va. Herp. Soc. Bull. no. 34:7.

*Witt, W. L. 1963c. Blue Ridge collecting notes. Va. Herp. Soc. Bull. no. 34:7.

*Witt, W. L. 1964a. Notes on Virginia reptiles. Va. Herp. Soc. Bull. no. 36:8.

*Witt, W. L. 1964b. Distribution of the snakes of Virginia. Va. Herp. Soc. Bull. no. 38:1–8.

*Witt, W. L. 1970. Opening a can of worm snakes? Va. Herp. Soc. Bull. no. 65:4.

*Witt, W. L. 1993. Annotated checklist of the amphibians and reptiles of Shenandoah National Park, Virginia. Catesbeiana 13 (2):26–35.

*Wood, J. T. 1954a. The distribution of poisonous snakes in Virginia. Va. J. Sci. 5 (3):152–167.

*Wood, J. T. 1954b. A survey of 200 cases of snake-bite in Virginia. Amer. J. Trop. Med. and Hygiene 3 (5):936–943.

*Wood, J. T., W. W. Hoback, and T. W. Green. 1955. Treatment of snake venom poisoning with ACTH and cortisone. Virginia Med. Month. 82:130–135.

*Wood, J. T., and R. H. Wilkinson. 1952. Size variations and sexual dimorphisms in a brood of common garter snakes, *Thamnophis o. ordinatus* (L). Va. J. Sci. 3 (3):202–205.

*Wood, J. T., and W. L. Witt. 1962. Herpetofauna collected by VHS members during first annual meeting at Camp Shawondasee, Chesterfield County, Virginia. Va. Herp. Soc. Bull. no. 30:3–4.

*Woolcott, W. S. 1959. Notes on the eggs and young of the scarlet snake, *Cemophora coccinea* Blumenbach. Copeia 1959 (3):263.

*Wright, A. H., and A. A. Wright. 1952. List of the snakes of the United States and Canada by states and provinces. Amer. Midl. Nat. 48 (3):574–603.

*Wright, A. H., and A. A. Wright. 1957. Handbook of snakes of the United States and Canada. 3 volumes. Ithaca, N.Y.: Comstock Publishing Associates.

*Wright, R. A. S. 1985. The ecology of probable venomous snake hibernacula habitat along the Blue Ridge Parkway, Bedford County, Virginia. Resource Management Project, Part One, National Park Service.

*Wright, R. A. S. 1987a. Natural history observations on venomous snakes near the Peaks of Otter, Bedford County, Virginia. Catesbeiana 7 (2):2–9.

*Wright, R. A. S. 1987b. Some important observations on the occurrence and frequency of venomous fauna found in illegal roadside solid waste accumulations in central Virginia. Va. J. Sci. 38 (2):126. Abstract.

*Wright, R. A. S. 1988a. Field notes: *Lampropeltis getulus getulus* (common or eastern kingsnake). Catesbeiana 8 (1):14.

*Wright, R. A. S. 1988b. Field notes: *Coluber constrictor constrictor* (black racer). Catesbeiana 8 (1):15.

*Wright, R. A. S. 1988c. Field notes: *Farancia erytrogramma* (rainbow snake). Catesbeiana 8 (1):15.

*Wright, R. A. S. 1988d. Field notes: *Heterodon platyrhinos* (eastern hognose snake). Catesbeiana 8 (1):15–16.

*Wright, R. A. S. 1988e. Field notes: *Thamnophis sirtalis sirtalis* (eastern garter snake). Catesbeiana 8 (2):31–32.

*Wright, R. A. S. 1990. Field notes: *Elaphe guttata guttata* (corn snake). Catesbeiana 10 (1):20.

*Wright, R. A. S. 1991. Field notes: *Carphophis amoenus* (worm snake). Catesbeiana 11 (1):19.

*Yarrow, H. C. 1882. Check list of North American Reptilia and Batrachia, with catalogue of specimens in U.S. National Museum. Bull. U.S. Nat. Mus. 24:1–249.

*Yongue, W. H., Jr. 1980. Mid-winter hematopoiesis in water snakes (genus: *Natrix*). Va. J. Sci. 31 (4):103. Abstract.

*Yongue, W. H., Jr. 1982. An apicomplexan (Protozoa) in the erythrocytes of the red belly water snake. Va. J. Sci. 33 (3):122. Abstract.

*Yongue, W. H., Jr., and K. A. Booker. 1979. The infectious agent of Cytotoddia (=*Toddia bufonis*). Va. J. Sci. 30 (2):53. Abstract.

*Young, D. A. 1993. An annotated checklist of reptiles and amphibians from Highland County, Virginia. Catesbeiana 13 (1):3–8.

*Zim, H. S., and H. M. Smith. 1953. Reptiles and amphibians. New York: Simon and Schuster.

INDEX